Advanced Methods for Modeling Water-Levels and Estimating Drawdowns with SeriesSEE, an Excel Add-In

By Keith Halford, C. Amanda Garcia, Joe Fenelon, and Benjamin Mirus

U. S. Department of Energy, National Nuclear Security Administration, Environmental Restoration Program, Underground Test Area Project

Techniques and Methods 4–F4

U.S. Department of the Interior
U.S. Geological Survey

U.S. Department of the Interior
KEN SALAZAR, Secretary

U.S. Geological Survey
Marcia K. McNutt, Director

U.S. Geological Survey, Reston, Virginia: 2012

For more information on the USGS—the Federal source for science about the Earth, its natural and living resources, natural hazards, and the environment, visit http://www.usgs.gov or call 1–888–ASK–USGS.

For an overview of USGS information products, including maps, imagery, and publications,
visit *http://www.usgs.gov/pubprod*.

To order this and other USGS information products, visit h*ttp://store.usgs.gov*.

Suggested citation:
Halford, K., Garcia, C.A., Fenelon, J., and Mirus, B., 2012, Advanced methods for modeling water-levels and estimating drawdowns with SeriesSEE, an Excel add-In, U.S. Geological Survey Techniques and Methods 4–F4, 28 p.

Preface

This report documents a spreadsheet add-in for viewing time series and modeling water levels that was developed in Microsoft® Excel 2010. Use of trade names does not constitute endorsement by the U.S. Geological Survey (USGS). The spreadsheet add-in has been tested for accuracy by using multiple datasets. If users find or suspect errors, please contact the USGS.

Every effort has been made by the USGS or the United States Government to ensure the spreadsheet add-in is error free. Even so, errors possibly exist in the spreadsheet add-in. The distribution of the spreadsheet add-in does not constitute any warranty by the USGS, and no responsibility is assumed by the USGS in connection therewith.

Acknowledgments

The report was prepared in cooperation with the U.S. Department of Energy, National Nuclear Security Administration, Nevada Site Office, Office of Environmental Management under Interagency Agreement DE-AI52-12NA30865.

The tide-challenged authors are indebted to Devin Galloway for clarifying the ebb and flow of tides.

Contents

Figures

Tables

Conversion Factors and Datums

Multiply	By	To obtain
Length		
foot (ft)	0.3048	meter (m)
mile (mi)	1.609	kilometer (km)
Volume		
gallon (gal)	3.785	liter (L)
Flow rate		
gallon per minute (gal/min)	0.06309	liter per second (L/s)
Transmissivity*		
foot squared per day (ft^2/d)	0.09290	meter squared per day (m^2/d)

Vertical coordinate information is referenced to the North American Vertical Datum of 1988 (NAVD 88).

Horizontal coordinate information is referenced to the North American Datum of 1983 (NAD 83).

Altitude, as used in this report, refers to distance above the vertical datum.

*Transmissivity: The standard unit for transmissivity is cubic foot per day per square foot times foot of aquifer thickness [(ft^3/d)/ft^2]ft. In this report, the mathematically reduced form, foot squared per day (ft^2/d), is used for convenience.

Advanced Methods for Modeling Water-Levels and Estimating Drawdowns with SeriesSEE, an Excel Add-In

By Keith Halford, C. Amanda Garcia, Joe Fenelon, and Benjamin Mirus

Abstract

Water-level modeling is used for multiple-well aquifer tests to reliably differentiate pumping responses from natural water-level changes in wells, or "environmental fluctuations." Synthetic water levels are created during water-level modeling and represent the summation of multiple component fluctuations, including those caused by environmental forcing and pumping. Pumping signals are modeled by transforming step-wise pumping records into water-level changes by using superimposed Theis functions. Water-levels can be modeled robustly with this Theis-transform approach because environmental fluctuations and pumping signals are simulated simultaneously. Water-level modeling with Theis transforms has been implemented in the program SeriesSEE, which is a Microsoft® Excel add-in. Moving average, Theis, pneumatic-lag, and gamma functions transform time series of measured values into water-level model components in SeriesSEE. Earth tides and step transforms are additional computed water-level model components. Water-level models are calibrated by minimizing a sum-of-squares objective function where singular value decomposition and Tikhonov regularization stabilize results. Drawdown estimates from a water-level model are the summation of all Theis transforms minus residual differences between synthetic and measured water levels. The accuracy of drawdown estimates is limited primarily by noise in the data sets, not the Theis-transform approach. Drawdowns much smaller than environmental fluctuations have been detected across major fault structures, at distances of more than 1 mile from the pumping well, and with limited pre-pumping and recovery data at sites across the United States. In addition to water-level modeling, utilities exist in SeriesSEE for viewing, cleaning, manipulating, and analyzing time-series data.

Introduction

Multiple-well, aquifer testing provides the most direct, integrated assessment of bulk hydraulic properties within complex geologic systems (Bohling and others, 2003; Sepúlveda, 2006; Yeh and Lee, 2007; Walton, 2008). The aquifer volume investigated with multi-well aquifer tests increases with increasing distance at which drawdown, or the pumping signal, can be detected (Risser and Bird, 2003; Halford and Yobbi, 2006). Drawdown analyses at distances of more than 1 mile (mi) often fail because environmental water-level fluctuations typically overwhelm the pumping signal. Barometric change, tidal forces, surface-water stage changes, or other external stresses induce these natural water-level changes in wells, which collectively are referred to here as "environmental fluctuations."

Barometric change and tidal forces can induce water-level fluctuations in a well greater than 1 foot (ft) during periods of less than a few days (Fenelon, 2000). Daily barometric changes alone typically exceed 0.3 ft where aquifers are confined or the unsaturated zone is thicker than 500 ft (Weeks, 1979; Merritt, 2004). Episodic recharge events can cause water-level rises that exceed 1 ft (O'Reilly, 1998). Climatic variations in recharge can induce long-term rising trends of more than 3 feet per year that affect detection of small pumping signals (Elliott and Fenelon, 2010; Fenelon, 2000). Drawdowns can be a fraction of the environmental fluctuations in distant observation wells that are more than a mile from a pumping well.

Environmental fluctuations have been modeled previously to differentiate natural water-level changes from pumping responses. Barometric and tidal effects typically are modeled independently and removed from water-level records (Erskine, 1991; Rasmussen and Crawford, 1997; Toll and Rasmussen, 2007). These approaches do not remove regional trends, such as long-term recharge, and are difficult to automate because all significant stresses that affect water levels other than pumping are not simulated simultaneously.

Water levels from background wells can be used to explicitly model water-level changes from recharge responses, surface-water stage changes, or any other external stress (Halford, 2006; Criss and Criss, 2011). A background well monitors water levels that are affected by tidal potential-rock interaction, imperfect barometric coupling, and all other stresses, excluding analyzed pumping, that affect water levels in observation wells. The need for antecedent data and background water levels has long been recognized (Stallman, 1971), but these trends and corrections typically have been estimated qualitatively.

Environmental fluctuations can be simulated as synthetic water levels, which represent the summation of multiple time series of barometric-pressure change, tidal potential, and background water levels, if available (Halford, 2006). Synthetic water levels are fitted to measured water levels for a period just prior to pumping, which should be more than three times greater than the period affected by pumping (Halford, 2006). Amplitude and phase of each time series are adjusted to minimize differences between synthetic and measured water levels. These synthetic water levels are projected into the pumping period, and drawdown is the difference between synthetic and measured water levels. This approach is referred to here as the "projection approach" to water-level modeling. The projection approach becomes unreliable where most of the analyzed period is affected by pumping.

Simultaneous modeling of environmental fluctuations and pumping signals overcomes the limitations of long-term extrapolation by using the projection approach. Environmental fluctuations can be defined during the entire period of record, which includes pumping and prolonged recovery periods. Variable pumping rates, as defined by a schedule of step changes, can be transformed to pumping signals by superimposing multiple Theis functions (Theis, 1935). Simultaneous simulation of all significant stresses affecting water-level changes is discussed as the "Theis-transform approach" to water-level modeling.

These water-level modeling approaches have been implemented in the program SeriesSEE, which is a Microsoft® Excel add-in. Water levels to be modeled, component fluctuations, and period of analysis are defined interactively and viewed in workbooks that are created by SeriesSEE. Water levels are modeled with a FORTRAN program that is called from Excel. Differences between synthetic and measured water levels are minimized with *PEST* (Doherty, 2010a and 2010b). Water-level models are calibrated rapidly because PEST files are created and executed seamlessly.

Water-level modeling with SeriesSEE differs from existing applications that filter environmental fluctuations or simulate pumping (Toll and Rasmussen, 2007; Harp and Vesselinov, 2011). This is because models of environmental fluctuations, Theis transforms, and parameter estimation are integrated in SeriesSEE. BETCO (barometric and earth tide correction) and similar programs simulate barometric and tidal water-level fluctuations but not regional trends and pumping effects (Toll and Rasmussen, 2007). Theis transforms have been applied previously in other water-level models, but environmental fluctuations were simulated with linear trends (Harp and Vesselinov, 2011).

Purpose and Scope

The purpose of this report is to document the approach used in SeriesSEE. This is the supporting software for modeling water levels that respond to environmental fluctuations and pumping. Water levels are modeled so pumping signals can be differentiated from environmental fluctuations. A method for fitting these water-level models to measured series by adjusting the selected parameters of each component is reported. The spreadsheet add-in is compatible with Microsoft® Excel 2010 (version 14.0) or higher. Use of the spreadsheet add-in requires basic knowledge of Excel. Use and applicability of this software is documented in this report. The hydrologic concepts and methods used in the data processing also are described briefly.

Environmental Fluctuations

Environmental fluctuations in measured water levels, or natural water-level changes, can be modeled by using pertinent time series, such as barometric pressure, tidal potential, background water levels, and stream stage. These time series represent potential components used to create synthetic water levels in a water-level model. Relevant components can be selected where a relation is expected with the water-level record. For example, water-level fluctuations in well b4mwh appear to be related to earth tide, barometric pressure fluctuations, recharge, and pumping (fig. 1). Simulating these environmental fluctuations in well b4mwh requires that earth tide, barometric pressure, and background water level (wells rw204 and sct4) components are included so that synthetic water levels can replicate measured water levels.

Barometric Effects

Barometric pressure induced water-level fluctuations are greatest in deep, confined aquifers where the rock matrix absorbs most of the atmospheric load (Merritt, 2004). Fluctuations increase because pressure instantly affects water levels in wells, whereas a stiff rock matrix transfers little of the increased atmospheric load to the confined water column. Atmospherically induced water-level fluctuations typically are less than 0.2 ft during a day. Large barometric-pressure changes from regional storms can cause water-level fluctuations of more than 1 ft during a week.

Barometric changes also measurably affect water levels in unconfined aquifers (Weeks, 1979). Pressure changes do not propagate instantaneously through the unsaturated zone because air is highly compressible. The relatively low pneumatic diffusivity of the unsaturated zone creates substantial phase lags between atmospheric and water-level changes. Unconfined water-level fluctuations can approach the magnitude of confined water-level fluctuations where the depth to water exceeds 500 ft. This is because atmospheric loading through the wellbore is not balanced by diffusion through the unsaturated zone.

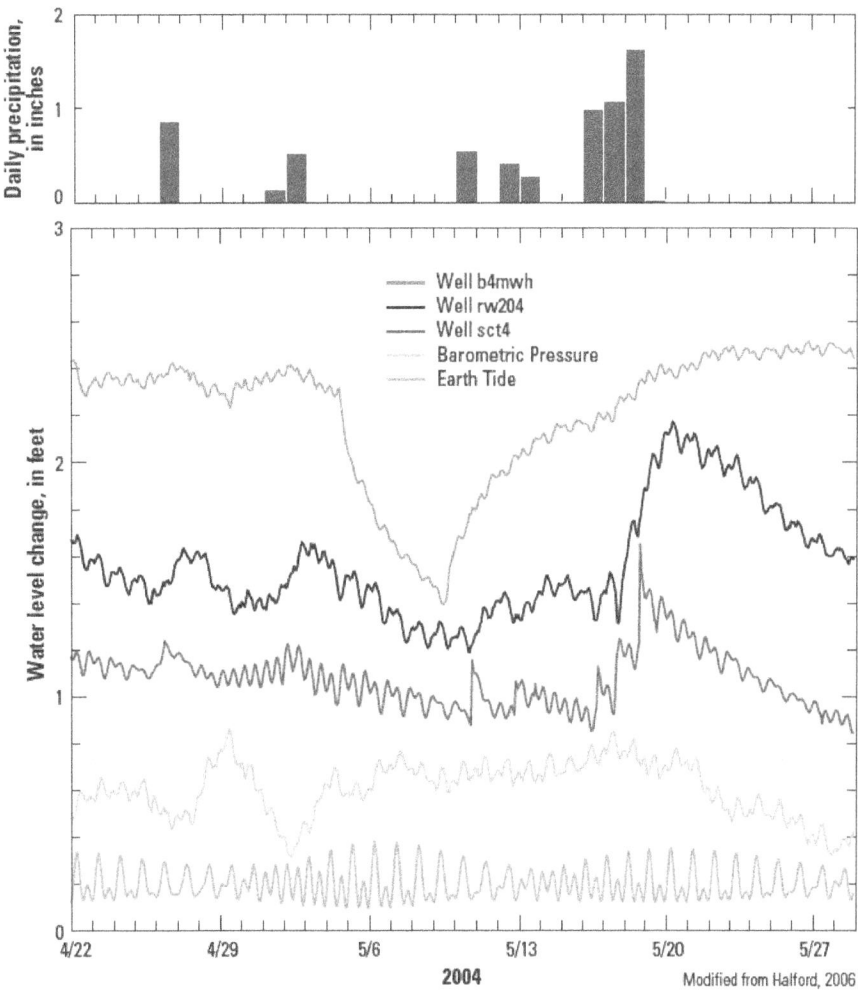

Figure 1.   Daily precipitation, groundwater levels, barometric change, and earth tide at Air Force Plant 6, Marietta, Georgia, April 22 to May 28, 2004.

Tidal Effects

Tidal forces distort the crust of the earth, which creates water-level fluctuations in mid-continent wells (Bredehoeft, 1967; Marine, 1975; Hanson and Owen, 1982; Narasimhan and others, 1984). Earth tides periodically deform (dilate and compress) the skeleton of the aquifer system, changing the porosity and causing measurable water-level fluctuations of as much as 0.1 ft or more in wells penetrating aquifers with small storage coefficients (fig. 1). Coupling between the mechanical deformation and the fluid filling the secondary porosity amplifies water-level response in wells hydraulically connected to the secondary-porosity features, such as fractures or faults. The presence of secondary porosity typically renders the formation more compliant to imposed stresses, depending on orientation of the fractures or faults with respect to the principal component directions of the imposed stress. The theoretical crustal strain tensors that result from the two principal lunar daily and semidiurnal tides are largely horizontal and orthogonal to one another. Subvertical fractures with azimuths oriented perpendicular to the strain tensor for a particular tide tend to amplify the strain and, thereby, the water-level response (Bower, 1983).

The diurnal rise and fall of ocean levels are the most common manifestation of varying gravitational forces and are referred to as ocean tides. Ocean tides affect coastal groundwater levels through direct head changes in an aquifer or as loads applied through a confining unit (Merritt, 2004). Ocean-tide effects are better approximated with a nearby tidal gage than calculated tides because wind and coastal geometry also affect ocean tides in addition to direct gravitational forcing.

Background Water Levels

Recharge events, regional pumping, and change in surface-water stage are identifiable stresses that typically affect large areas but are not predicted easily with independent time series such as barometric change and tidal potential. Recharge events and regional pumping stresses can create similar water-level changes in multiple wells over areas of many square miles. Change in surface-water stages locally affects groundwater levels and can be measured directly. Water levels in wells sufficiently removed from an aquifer test can simulate these regional stresses, local changes in surface-water stages, and any other unidentified pervasive stresses. Water levels in these remote wells are referred to as background water levels (Halford, 2006).

Background water levels can be more effective correctors than independent barometric and tidal time series even where only barometric and tidal stresses are significant (Halford, 2006). Barometric forcing through the unsaturated zone lags behind water-level changes because of the small permeability of unsaturated rock relative to an open well (Weeks, 1979). The complex relation between barometric pressure and water level in a well is explained poorly with barometric efficiency where the unsaturated zone is thick. Background water levels from another well of similar construction better approximate this relation. Likewise, rock properties and fracture orientation in an aquifer control tidal water-level fluctuations as much as tidal forcing. Water levels from background wells can better approximate the rock-tide interaction than theoretical tidal components alone. Independent barometric and tidal time series frequently remain necessary because of differences in rock properties, fracture orientation, and well completions around measured and background wells.

Water-Level Modeling

Water-level modeling assumes that measured water-level fluctuations can be approximated by summing multiple-component fluctuations (Halford, 2006). Input series of barometric pressure, input series of background water levels, and computed earth tides explain most environmental fluctuations (fig. 2). Pumping signals are simulated with multiple Theis solutions that transform pumping schedules to water-level fluctuations.

Water-level model components are summed to create a synthetic water level. A synthetic water level at time, t, is determined:

$$SWL(t) = C_0 + \sum_{i=1}^{n} WLMC_i \qquad (1)$$

where

$\quad C_0 \quad$ is an offset (L) that allows mean values of synthetic water levels to match mean values of measured water levels,

$\quad n \quad$ is the number of water-level model (WLM) components, and

$\quad WLMC_i \quad$ is the ith WLM component in units of the modeled water level.

Water-level model results are denoted with the word synthetic rather than simulated to differentiate between water-level and groundwater-flow model results.

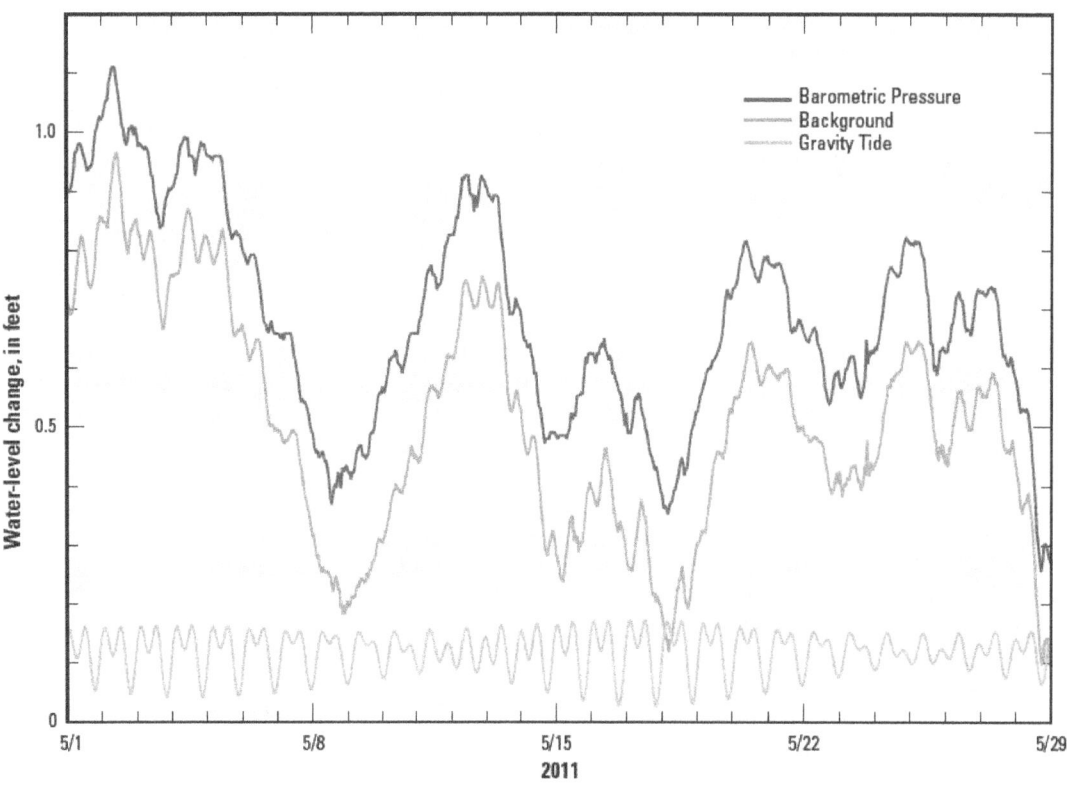

Figure 2.   Input series of barometric pressure, input series of background water level, and computed gravity tide.

Water-Level Model Components

Input series are measured water levels, barometric pressures, or pumping schedules that are transformed to represent water-level change. All input series are assumed to be continuous between each discrete measurement where continuity can be piecewise linear or stepwise. Water levels and barometric pressures typically are used as piecewise linear functions. Pumping schedules typically are used as stepwise functions. All input series are transformed into WLM components that are smooth, differentiable functions.

WLM components are created from input series with one of six transforms. The parameters that define each transform generically are referred to as coefficients because characteristics and terminology are not consistent among transforms (table 1). Moving averages are most frequently used to transform interpolated time series of barometric pressure and background water levels into WLM components. Pumping schedules are transformed into water-level fluctuations with Theis transforms. Earth tides are computed for a given observation well location (Harrison, 1971). Transducer displacement, as a result of resetting a transducer in a well, is simulated with the step transform following a user-specified time. Lag and attenuation of barometric-pressure changes between land surface and water table are simulated with the pneumatic-lag transform. Water-level rises from infiltration events are simulated with the gamma transform.

WLM components are smooth functions because values are interpolated linearly between consecutive data pairs or transformed from stepwise data to a smooth function. Interpolation or transformation allows data to be collected at variable intervals within a time series. Collection frequencies can differ among time series and do not need to be synchronized because interpolation or transformation synchronizes comparisons (fig. 3).

Moving Average

Fluctuations of different frequencies exist in input series such as barometric changes and background water levels. Barometric changes exhibit diurnal, weekly, and seasonal fluctuations that differ in amplitude and frequency. Frequency-dependent differences in water-level fluctuations also exist between wells because of differences in well construction and aquifer properties. Diurnal water-level fluctuations will be less where communication between well and aquifer is impeded and wellbore storage is increased. Poorly developed wells with large casing diameters and short screens damp high-frequency water-level fluctuations. Aquifers with large storage coefficients and small transmissivity values also will damp water-level fluctuations.

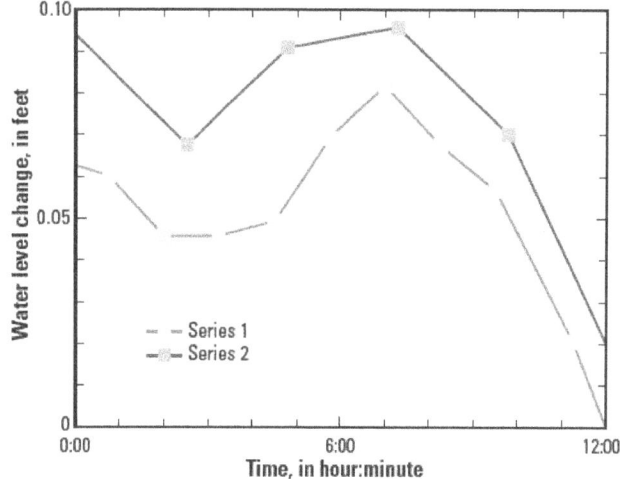

Figure 3. Two time series with different collection frequencies and sampling times.

Table 1. Water-level model (WLM) components.

[— is not applicable]

WLM component	Time series	Coefficient				
		1	2	3	4	5
Moving average	Any series	Multiplier	Phase	Averaging period	—	—
Theis transform	Pumping schedule	Transmissivity	Storage coefficient	Radial distance	Flow-rate conversion	—
Tide	Computed	Multiplier	Phase	Latitude	Longitude	Altitude
Step	—	Time	Offset	—	—	—
Pneumatic lag [a]	Barometric pressure	K_{AIR}	S_{AIR}	Thickness of unsaturated zone	—	—
Gamma[1]	Infiltration	Multiplier	k	n	Time conversion	Multiplication series

[a] Hydraulic properties of the Pneumatic-lag transform, K_{AIR} & S_{AIR}, are with respect to air. K_{AIR} is hydraulic conductivity of air and is about 60 times greater than K_{WATER}. S_{AIR} is average air-filled porosity divided by mean air pressure.

[1] The k and n terms represent scale and shape parameters, respectively in the Gamma Probability Distribution Function.

Input series frequently are composed of multiple signals of different frequencies. These different frequencies can be separated into multiple WLM components with multiple moving averages of the input series (fig. 4). Water levels can be averaged over periods of hours to days where duration of averaging periods and the number of WLM components are arbitrary quantities. More than a half dozen WLM components frequently are created from a single input series because a broad range of averaging periods are more likely to simulate the environmental fluctuations. An excess of WLM components generally does not degrade results. High-frequency signals are approximated indirectly by summing multiple WLM components with ranges of averaging periods. The original input series and WLM component are one and the same where an averaging period of 0 is specified (table 1).

The moving-average transform is applied to i^{th} WLM component at time, t:

$$WLMC_i = a_i V_i(t + \phi_i) \qquad (2)$$

where

a_i is the amplitude multiplier of the ith component in units of the modeled water level divided by units of the ith component,

Φ_i is the phase-shift of the ith component (t), and

$V_i(t+\Phi_i)$ is the value of the moving average of ith input series at time $t + \Phi_i$ in units of ith component.

Amplitude (a) and phase (Φ) are estimated in equation 2 to minimize differences between synthetic and measured water-levels.

Moving averages are centered about the evaluation time, t, where averaging periods are defined by time, not the number of measurements. For example, a 12-hr, moving average at the time when sampling increased from hourly to 15-minute measurements would average 31 values. Six values were measured prior to the evaluation time, another value was measured at the evaluation time, and 24 values were measured after the evaluation time.

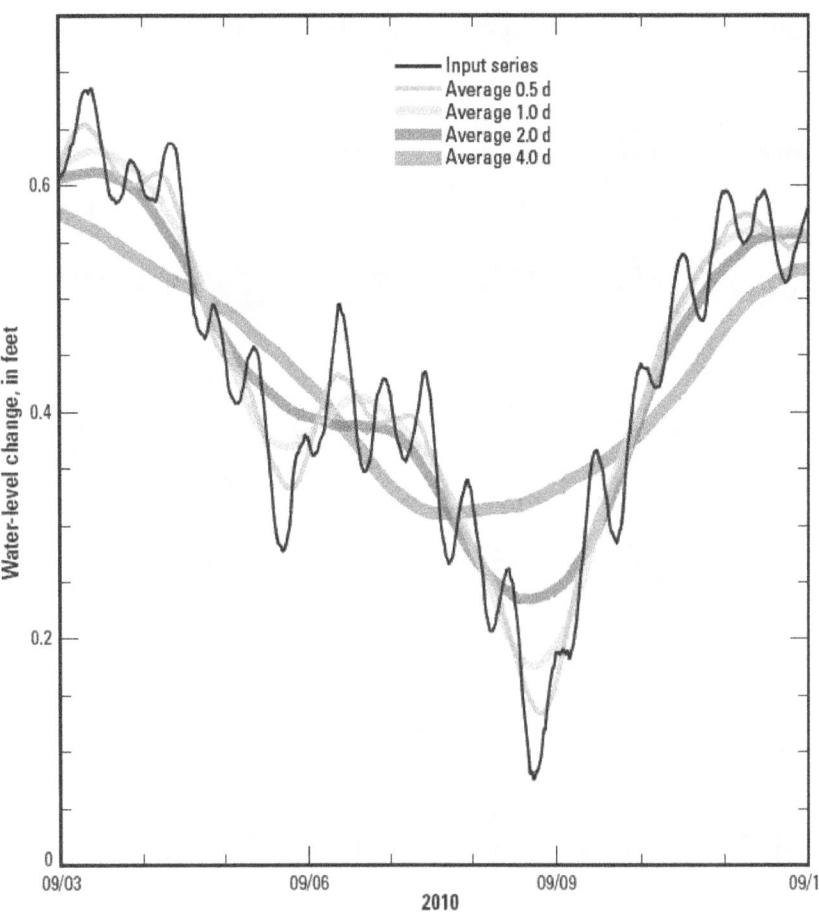

Figure 4.   Input series and four additional water-level model components that were created by averaging in periods of 0.5, 1, 2, and 4 days (d).

Theis Transform

Pumping schedules are converted into water-level responses with a simple model: the Theis (1935) solution. Water-level changes or drawdown, s, from pumping-rate changes are simulated:

$$WLMC_i = s = \frac{Q}{4\pi T}W(u) = \frac{Q}{4\pi T}W\left(\frac{r^2 S}{4T\Delta t}\right) \qquad (3)$$

where

Q	is the flow rate (L³/t),
T	is the transmissivity (L²/t),
$W(u)$	is the exponential integral solution,
u	is dimensionless time,
r	is the radius (L),
S	is the storage coefficient (dimensionless), and
Δt	is the elapsed time since the flow rate changed (t).

Multiple Theis solutions are superimposed in time to simulate water-level responses to variable pumping schedules (fig. 5). The effects of multiple pumping wells also can be simulated by superposition in space (Harp and Vesselinov, 2011). Each pumping well with its unique pumping schedule and radial distance is simulated with a WLM component in SeriesSEE. Pumping signals are discussed here as drawdowns, regardless of pumping rate, because discrete drawdown and recovery periods do not exist when variable pumping schedules are simulated.

Superimposed Theis solutions serve as transform functions, where step-wise pumping records are translated into approximate water-level responses at observation wells. Log-transforms of transmissivity (T) and storage coefficient (S) are estimated in equation 3 to minimize differences between synthetic and measured water-levels. Estimates of T and S can characterize correctly the hydraulic properties of an aquifer if assumptions of the Theis solution are honored. These same

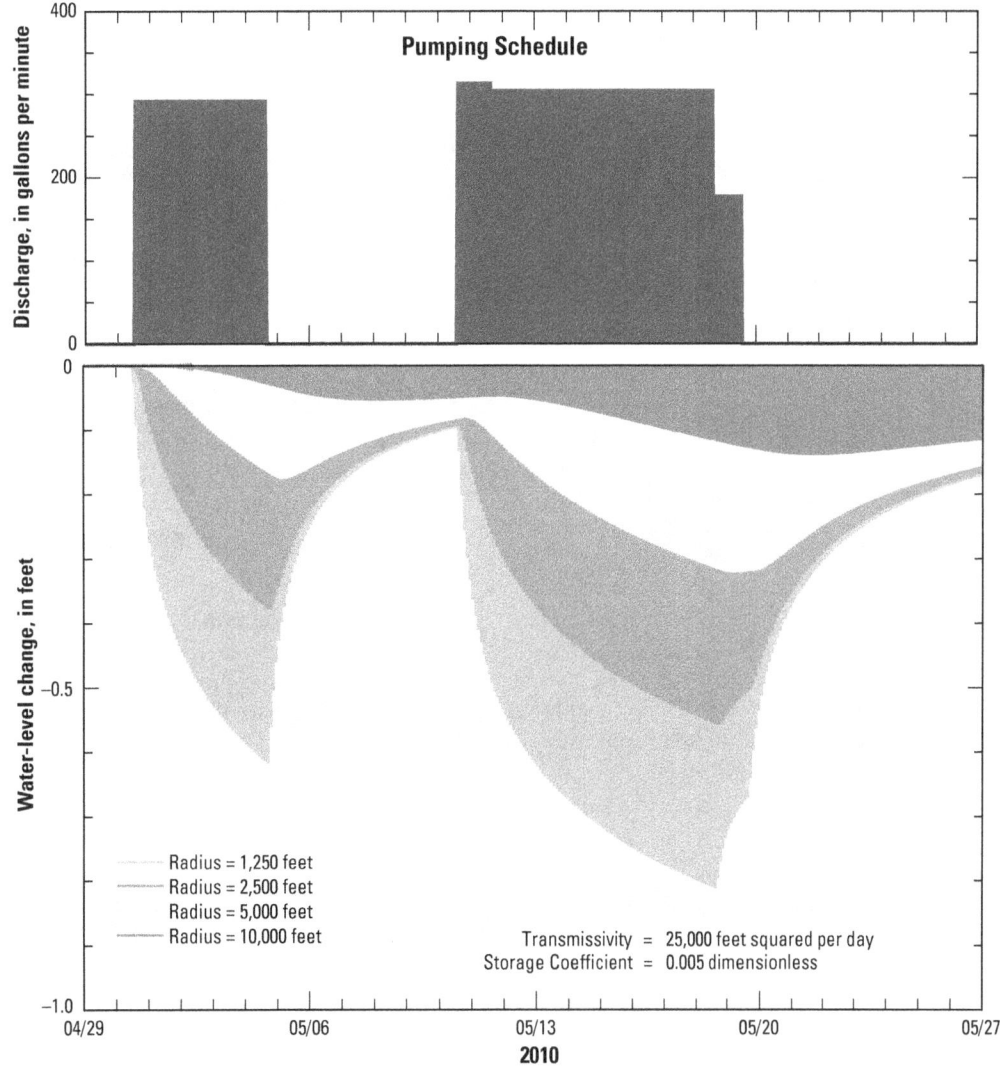

Figure 5. Theis transform of a pumping schedule to water-level changes at radial distances between 1,250 and 10,000 feet from a pumping well for a fixed transmissivity and storage coefficient.

parameters primarily are fitting terms with little physical significance in hydrogeologically complex aquifer systems because assumptions of the Theis solution are violated. This component of the water-level model is referred to as a "Theis transform," here, and applies to the pumping schedule of a single well.

Hydrogeologic complexity and uncertainty are addressed by applying multiple Theis transforms to a single pumping schedule. Relatively fast and slow elements of pumping signals propagate through complex aquifer systems. These fast and slow elements are approximated by Theis transforms with relatively high and low hydraulic diffusivities, respectively.

Computed Tides

The tides are displacements of the particles in a celestial body caused by the forces of attraction in a neighboring body. The terrestrial tides on Earth consist of the atmospheric tides, the earth tides, and the ocean tides and are related to the lunar and solar cycles (Defant, 1958). Simulated tidal forcing and body tides of a solid Earth (oceanless) produced by the moon and sun are computed from gravitational and astronomical theory for a specified point on the Earth for a specified time by using the Harrison (1971) model. Changes in the solid Earth caused by the ocean tides are not considered here. Many of the model parameters, and thus the computed tidal components, are functions of time based on the ephemerides, which are computed in the model but are not included here explicitly.

The earth tides result as the crust undergoes volumetric strains, ε_V, due to variations in tide-generating forces:

$$\varepsilon_V = \frac{1}{3}\left(\varepsilon_{\theta\theta} + \varepsilon_{\lambda\lambda} + \varepsilon_{rr}\right) \qquad (4)$$

where, $\varepsilon_{\theta\theta}$, $\varepsilon_{\lambda\lambda}$, and ε_{rr} (positive downwards) represent the principal components of the strain-tide tensor with respect to polar north, east, and radial, respectively. Most of the stress close to the Earth's surface is plane stress, and the resultant strain tide is predominately an areal strain, ε_A (Melchior, 1966:

$$\varepsilon_A = \frac{1}{2}\left(\varepsilon_{\theta\theta} + \varepsilon_{\lambda\lambda}\right) \qquad (5)$$

The areal strain produced by earth tides is computed from theoretical considerations (Harrison, 1971, 1985; Beaumont and Berger, 1975; Berger and Beaumont, 1976) by using the tidal potential, V (L²/t²), as formulated by Bartels (1957, 1985) and computed by Harrison (1971):

The areal strain tide component is formulated as a scaled function of the tidal potential (Munk and McDonald, 1960; Melchior, 1966, Bredehoeft, 1967):

$$\varepsilon_A = \left(2\bar{h} - 6\bar{l}\right)\frac{V}{rg} \qquad (7)$$

where

$\bar{h}$ and $\bar{l}$ are Love numbers at the Earth's surface, and

g is the gravitational acceleration (L/t²).

Areal strain tide is computed by using $\bar{h} = 0.638$ and $\bar{l} = 0.088$ and is expressed in parts per billion strain (dimensionless). The resulting areal 'dry' (in the absence of saturating fluid) tidal dilatation at the Earth's surface, Δ_t can be expressed (Bredehoeft, 1967):

$$\Delta_t = \left[\frac{1 - 2v}{1 - v}\right]\varepsilon_A \qquad (8)$$

where v is Poisson's ratio.

The gravity tide oriented downwards normal to the Earth's ellipsoid, g_N, is computed (Harrison, 1971):

$$g_N = \frac{\partial V}{\partial r} - \delta \frac{\partial V}{r\partial\theta} \qquad (9)$$

where

θ is the geocentric polar angle of the observation point (radians), and

δ is the difference between the geodetic and geocentric latitudes.

For example, δ attains a value of about 3.37×10^{-3} radians at 45° latitude. Gravity tide is expressed in terms of microgals (L/t²).

The tilt tide in a plane tangent to the Earth's ellipsoid along a specified azimuth oriented with respect to 0° N, γ_T is computed (Harrison, 1971):

$$\gamma_T = \frac{1}{g}\left[\left(\frac{\partial V}{r\partial\theta} + \delta\frac{\partial V}{\partial r}\right)\cos\alpha + \frac{1}{r\sin\theta}\frac{\partial V}{\partial\lambda}\sin\alpha\right] \qquad (10)$$

where

λ is the terrestrial east longitude of the observation point (radians) and

α is the specified azimuth of tilt (radians).

Tilt tide is expressed in nanoradians.

$$V = \frac{GMr^2}{R_s^3}\left\{\frac{3\cos^2 z_m - 1}{2} + \frac{r}{R_m}\frac{5\cos^2 z_m - 3\cos^2 z_m}{2}\right\} + \frac{GSr^2}{R_s^3}\left\{\frac{3\cos^2 z_s - 1}{2}\right\} \qquad (6)$$

where

G is the Newtonian constant of gravitation (L³ / M¹-t²),

M and S are the masses of the moon and sun, respectively (M),

r is the distance between the center of the Earth and the observation point on the Earth's surface (L),

R_m and R_s are the distances of the moon and sun, respectively, from the Earth's center (L), and

z_m and z_s are the zenith angles of the moon and sun, respectively (radians).

Dry, gravity, and tilt tides (Table 2) result from changes in gravitational forces as the relative positions of the sun, moon, and earth change (Harrison, 1971). These theoretical earth tides are computed functions that only require the location of an observation well.

Adjustable WLM components are created by multiplying computed dry, gravity, or tilt tide (table 2) by an amplitude. Zenith angles primarily are specified by longitude and time as referenced to Greenwich Mean Time. A phase shift can be applied to the zenith angles through the specified time. Amplitude (a) and phase (Φ) are estimated to minimize differences between synthetic and measured water-levels.

Step Change

Step changes in water-level records are introduced when a transducer is disturbed or replaced. Transducer submergence can change if the hanger position is moved. Replacing a transducer is likely to change submergence because the devices can differ and cable stretch can occur. A step-change WLM component is necessary because shifts of less than 0.03 ft are detectable in WLM results.

A step change in the water-level measurement is simulated as follows:

$$WLMC_i = \Delta h_i \qquad \text{for } t \geq t_{STEP}$$
$$WLMC_i = 0 \qquad \text{for } t < t_{STEP}$$

(11)

where

Δh_i is the step change of the ith component and t is the time. The step change is estimated in equation 11 to minimize differences between synthetic and measured water-levels.

Pneumatic Lag

The pneumatic lag between barometric-pressure changes at land surface and the water table can be simulated with a one-dimensional diffusion equation instead of being approximated with multiple moving averages. This alternative approach is advantageous for estimating the hydraulic properties of the unsaturated zone and precludes using multiple moving averages of barometric pressure. The propagation of barometric changes through the unsaturated zone is solved analytically by using equivalent solutions for surface-water/groundwater interaction (Rorabaugh, 1964; Barlow and Moench, 1998).

Stage changes of a fully penetrating river that perturb groundwater levels behave similarly to barometric pressure changes that perturb air pressures in the unsaturated zone (fig. 6). This assumes that pressure changes are small relative to the mean air-pressure so air density and specific storage are affected minimally. Barometric changes typically are less than 2 ft while mean air-pressure ranges between 26 and 34 ft (Merritt, 2004; Fenelon, 2005). Boundary conditions for a one-dimensional, confined aquifer are equivalent to boundary conditions of an areally extensive, thick unsaturated zone. The water table is an impermeable boundary because air-filled pores cease to exist.

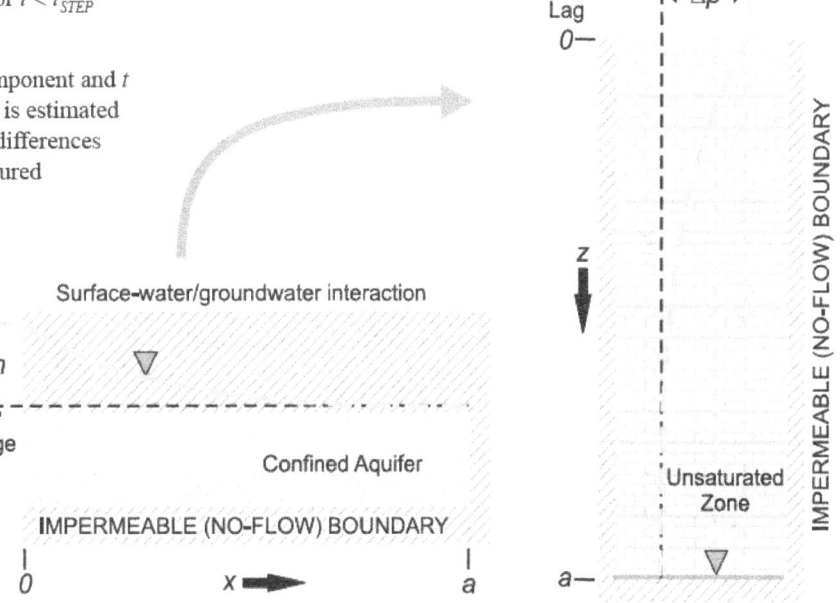

Figure 6. Schematics of one-dimensional, confined aquifer and an areally extensive, thick unsaturated zone that experience similar step-changes to a time-varying specified-head boundary such as a river or barometric-pressure difference.

Table 2. Abbreviations and descriptions of tides that are computed in SeriesSEE.

Tide	DESCRIPTION	Units	Equation
DRY	Areal strain tide	parts per billion	8
GRAVITY	Normal to the Earth ellipsoid	microgals	9
TILT	Plane tangent to the Earth ellipsoid	nanoradians	10

Equivalent hydraulic conductivity and specific storage of the unsaturated zone differ from the confined aquifer solution because the pores are filled with air rather than water. Equivalent hydraulic conductivity is air permeability divided by the viscosity of air and is about 60 times greater than saturated hydraulic conductivity because the ratio of water-to-air viscosity ranges from 70 to 40 for temperatures between 10 and 30°C. Air permeability is affected negligibly by changes in barometric pressure (Baehr and Hult, 1991). Specific storage of the unsaturated zone is the air-filled porosity divided by the mean air pressure.

Pressure change at a given depth in the unsaturated zone from a step-change in pressure at land surface is simulated as follows:

Water-table changes are assumed equal and opposite of air-pressure changes at the water table. Log-transforms of K_{AIR} and S_{AIR} are estimated in equation 12 to minimize differences between synthetic and measured water-levels. If the objective of a water-level model is to estimate hydraulic properties of the unsaturated zone by using equation 11, then multiple moving averages of barometric pressure cannot be used as WLM components.

$$WLMC_i = \Delta p - 2 \sum_{m=1}^{\infty} \frac{-1^{m+1}}{\left(\pi m - \frac{1}{2}\right)} e^{-\pi^2 \left(m - \frac{1}{2}\right)^2 \frac{\Delta t K_{AIR}}{S_{AIR} a^2}} \cos\left(\pi\left(m - \frac{1}{2}\right)\left(1 - \frac{z}{a}\right)\right) \qquad (12)$$

where

Δp	is the step change in air pressure at land surface (L),
m	is an index,
Δt	is elapsed time since the step change (t),
K_{AIR}	is the air permeability divided by viscosity of air (L/t),
S_{AIR}	is air-filled porosity divided by the mean air-pressure (1/L), and
a	is the thickness of the unsaturated zone (L).

Multiple step changes are superimposed in time to simulate air-pressure changes at the water table by using barometric-pressure changes at land surface (fig. 7).

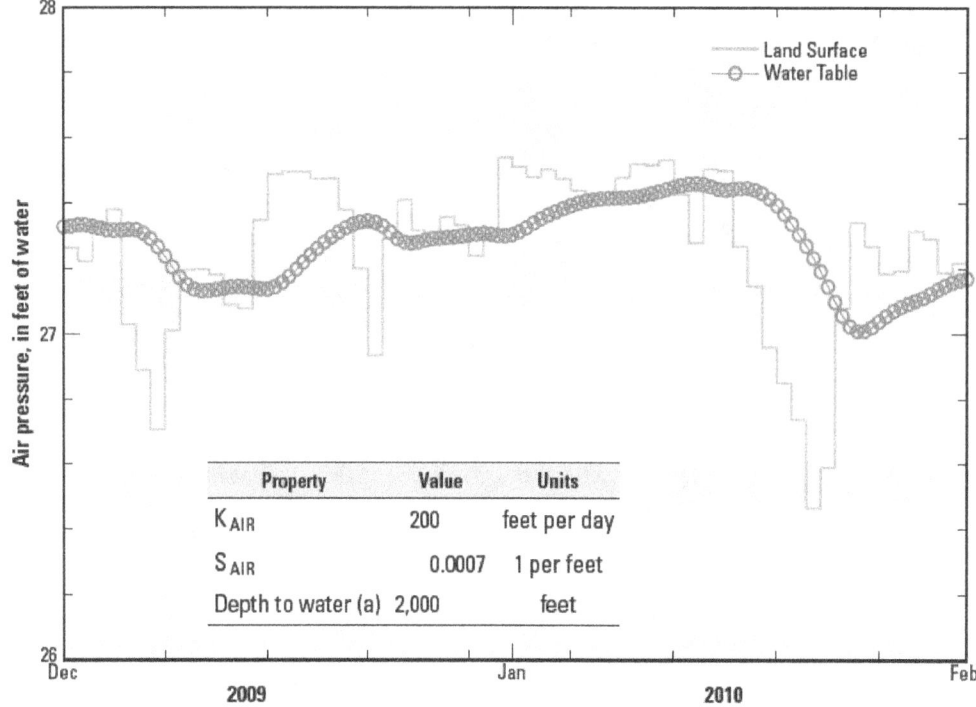

Figure 7.   Average daily barometric pressure and simulated air pressure at the water table.

Gamma Transform

The gamma transform was adapted from a Water-Balance/Transfer Function (WBTF) model that simulates recharge to the water table from precipitation (O'Reilly, 2004). The gamma transform retains the transfer function from the WBTF model that translates a discrete pulse of infiltration below the root zone to recharge at the water table. The delay between infiltration and recharge at the water table increases as unsaturated-zone thickness increases. Recharge pulses also are attenuated and prolonged as unsaturated-zone thickness increases. The WBTF model was selected because the transfer function simulates these characteristics (O'Reilly, 2004).

Water-level rise, rather than recharge, is simulated with the gamma transform. Water-level rise equals recharge divided by specific yield, where the aquifer is unconfined, and consequently has a greater magnitude than recharge (fig. 8).

Water-table rise from each infiltration event is simulated as follows:

$$WLMC_i = a_i I \frac{e^{-\frac{\Delta t}{k}}}{k\Gamma(n)}\left(\frac{\Delta t}{k}\right)^{n-1} \tag{13}$$

where

a_i	is the amplitude multiplier of the ith component,
I	is amount of infiltration during an event (L),
Δt	is elapsed time since the infiltration event(t),
k	is a scale parameter (t),
n	is a shape parameter (dimensionless) , and
$\Gamma(n)$	is the gamma function, (dimensionless), which is equivalent to $(n-1)$ for integer values of n (Potter and Goldberg, 1987, p. 111).

Multiple step changes are superimposed in time to simulate water-table fluctuations from infiltration events below land surface (O'Reilly, 2004).

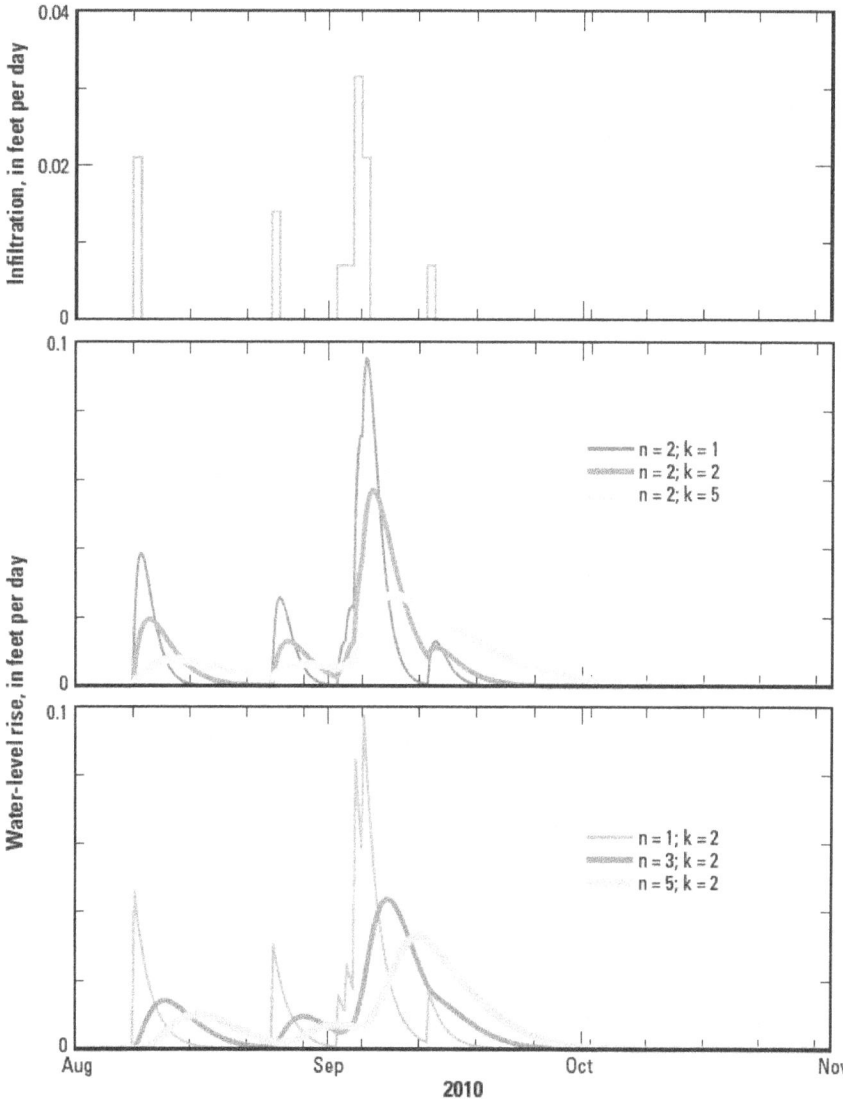

Figure 8. An infiltration schedule and water-level rises simulated with gamma transforms that were defined by six pairs of shape (n) and scale (k) parameters.

Physical significances have been attributed to the fitting parameters a_i, k, and n (O'Reilly, 2004). The amplitude multiplier (a_i) converts recharge to water-level rise and should be proportional to the inverse of the storage coefficient. The scale parameter (k) controls the average delay time imposed by the unsaturated zone (Dooge, 1959). The shape parameter (n) has been characterized as "the number of linear reservoirs necessary to represent the unsaturated zone" by O'Reilly (2004). These explanations are interesting, but estimated values of a_i, k, and n should be interpreted with great skepticism, if at all.

Superimposed gamma transforms translate step-wise precipitation or infiltration records into approximate water-level responses at observation wells. Amplitude (a) and the log-transform of the scale parameter (k) are estimated in equation 13 to minimize differences between synthetic and measured water-levels. The shape parameter (n) is assigned and is not estimated. Multiple gamma transforms should be used with different values of n if the effect of n is investigated.

Calibration

Water-level models must be calibrated to reliably differentiate small pumping responses from environmental fluctuations. Efficient and effective calibration requires a quantitative measure of model misfit so model parameters can be estimated automatically as is done with the parameter estimation software PEST (Doherty, 2010a, 2010b). Differences between synthetic and measured water levels, or residuals, define the goodness-of-fit and are summed in the measurement objective function:

$$\Phi(x)_{MEAS} = \sum_{j=1}^{nobs} \left(SWL(x)_j - MWL_j \right)^2 \qquad (14)$$

where

x	is the vector of parameters being estimated,
$nobs$	is the number of observations compared,
$SWL(x)_j$	is the jth synthetic water level, and
MWL_j	is the jth measured water level.

Although the sum-of-squares error serves as the measurement objective function, root-mean-square (RMS) error,

$$RMS = \sqrt{\frac{\Phi(x)_{MEAS}}{nobs}} \qquad (15)$$

is reported because RMS is easily compared to measurements.

Residuals are not weighted in the measurement objective function because suspect measured water levels should be discarded rather than assigned a low weight. Each measured water level is assumed equally important so all water levels are weighted equally. Uniform weighting causes differences between synthetic and measured water levels to equally affect the measurement objective function (eq. 14).

Stable parameter-estimation results are ensured with selective parameter transformation and regularization. Log-transforms of hydraulic properties are estimated in the Theis, pneumatic lag, and gamma transforms to scale parameters and precluded negative hydraulic properties (table 3). Regularization avoids estimating insensitive parameters and guides estimates toward preferred values. Parameter estimates have little to no significance because the parameter values generally are not interpreted. Drawdown estimates are interpreted and are the ultimate water-level model result.

Parameter estimation for water-level modeling is unconditionally stable because singular-value decomposition (SVD) regularization is used (Doherty and Hunt, 2010). Insensitive or highly correlated parameters are not estimated and remain at their assigned values if eliminated by SVD regularization.

Tikhonov regularization guides estimates to preferred conditions (Doherty, 2010a, 2010b). Regularization observations are added to define preferred relations between parameters (Doherty and Johnston, 2003). Homogeneity within each of the three parameter groups of amplitude, phase, and hydraulic property was the preferred relation that was enforced with Tikhonov regularization (table 3).

The balance between fitting measurement and regularization observations is controlled by the sum-of-squares measurement error, PHIMLIM, in PEST (Doherty, 2010a, 2010b). An expected RMS error defines PHIMLIM, which equals the square of the expected RMS error times the number of measured water levels (*nobs*). The expected RMS error defaults to 0.003 (L) in SeriesSEE, but can be changed by the user.

Table 3.   Summary of estimable parameters and parameter groups for water-level modeling (WLM) components.

[— is not applicable]

WLM component	Coefficient 1	Parameter group	Coefficient 2	Parameter group
Moving Average	a	Amplitude	φ	Phase
Theis Transform	T	Hydraulic Property	S	Hydraulic Property
Tide	a	Amplitude	φ	Phase
Step	—	—	a	Amplitude
Pneumatic Lag	K_{AIR}	Hydraulic Property	S_{AIR}	Hydraulic Property
Gamma	a	Amplitude	k	Hydraulic Property

Drawdown Estimation

Drawdown estimates from a water-level model are the difference between measured water levels and synthetic water levels without the Theis transforms. Alternatively, drawdowns can be computed directly by summing all Theis transforms and subtracting residuals (fig. 9). The summation of all Theis transforms is the direct estimate of the pumping signal. Residuals represent all unexplained water-level fluctuations. These fluctuations should be random residuals during non-pumping periods, but can contain unexplained components of the pumping signal during pumping and recovery periods. This method of estimating drawdowns is called the Theis-transform approach.

A limited, application of water-level modeling, the projection approach, was developed prior to the Theis-transform approach (Halford, 2006). Synthetic water levels were developed and calibrated during a period prior to pumping with the projection approach. Calibrated, synthetic water levels were then projected forward during pumping and recovery. Drawdown was the difference between projected synthetic values and measured values. This approach ensures that environmental fluctuations and the pumping signal are uncorrelated because pumping is not simulated during model calibration to antecedent water levels.

The projection approach is limited primarily because regional water-level trends are simulated poorly. Excluding pumping and recovery periods from WLM calibration eliminated much of the regional trends from the calibration period. This drawback weakened the projection approach and limited the usefulness of background well information, particularly where pumping and recovery periods were greater than the antecedent data period.

The Theis-transform approach is a more robust application of water-level modeling because environmental fluctuations and pumping signal are simulated during pumping and recovery in addition to antecedent water levels. This allows for calibration of synthetic water-levels to all measured data. The effects of pumping on measured water levels are approximated by using a simple approach, Theis transforms, so that simulations are quick. Efficiency and speed are mandatory because water levels are modeled independently in every observation well. These requirements preclude numerical groundwater-flow models or any other laborious approach for translating pumping schedules to water-level responses.

Drawdown detection with the Theis-transform approach becomes ambiguous when the signal-to-noise ratio is low or where environmental fluctuations and pumping signals can be correlated. Signal and noise are defined herein as the maximum drawdown in a well during an aquifer test and the RMS

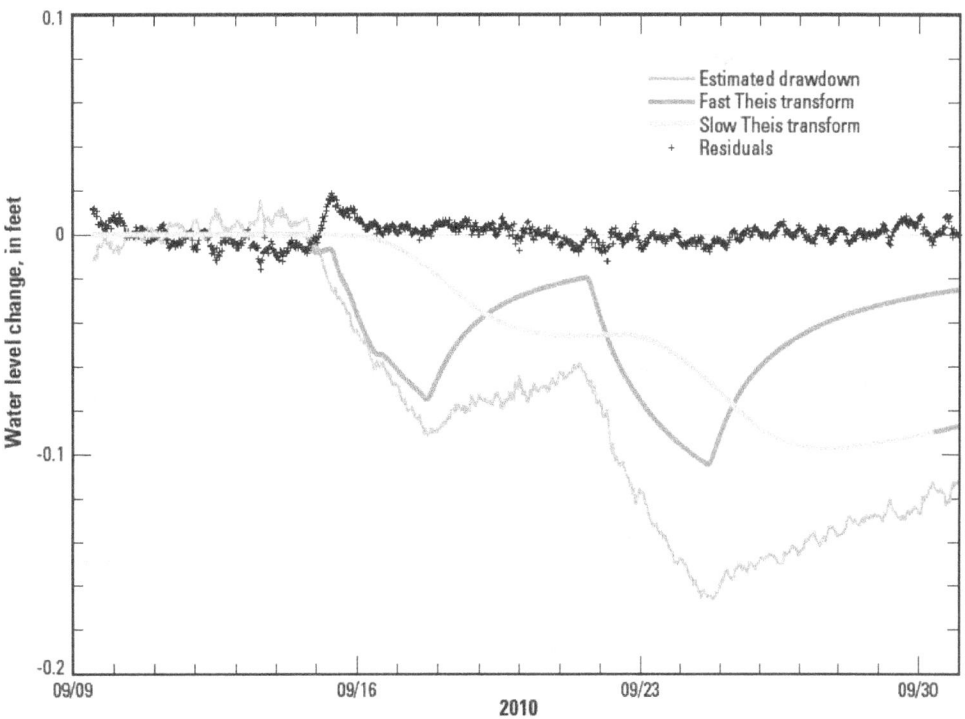

Figure 9. Estimated drawdown from summing Theis transforms and subtracting residuals. Fast and slow Theis transforms represent the relatively fast and slow elements of pumping signals that propagate through a complex aquifer system.

error, respectively. Drawdown has been detected definitively where the signal-to-noise ratio was greater than 10 and correlation was unlikely. Correlation is unlikely where sharply defined pumping signals (saw-tooth shape) exist or considerable recovery has been observed (fig. 10, ER-EC-6 deep, r = 6,800 ft). Correlation between environmental fluctuations and the pumping signal is possible where observed drawdown can be approximated by a linear trend during all or part of the period of analysis (fig. 10, ER-EC-12 shallow, r = 8,900 ft). The potential for correlation increases as hydraulic diffusivity decreases, distance between observation and pumping well increases, or recovery diminishes.

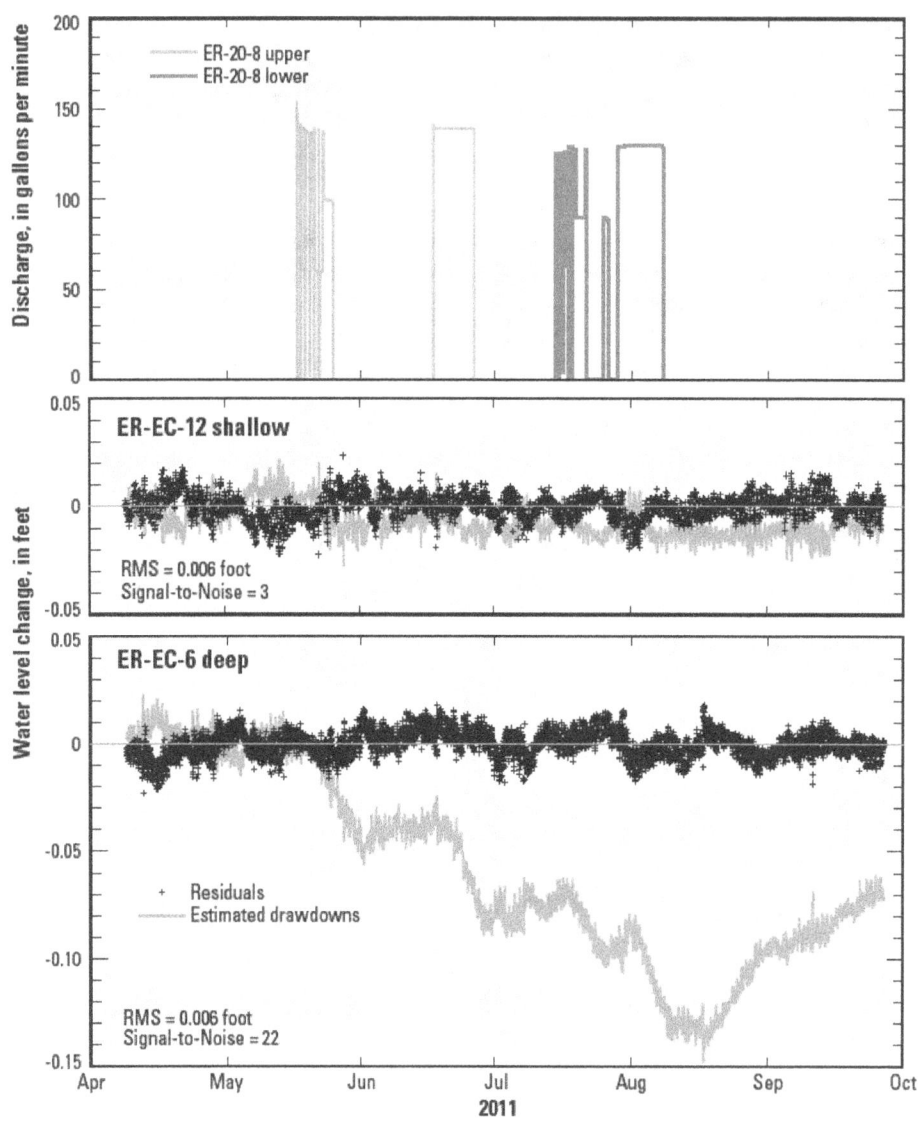

Figure 10.   Discharge from pumping wells ER-20-8 upper and ER-20-8 lower, estimated drawdowns, residuals, RMS errors, and signal-to-noise ratios in observation wells ER-EC-12 shallow and ER-EC-6 deep.

SeriesSEE

SeriesSEE is a Microsoft® Excel add-in for viewing, cleaning, manipulating, and analyzing time-series data where water-level modeling is a primary analysis tool. SeriesSEE creates a viewer file from a data workbook that can contain more than 16,000 series. The maximum number of series that can be viewed simultaneously is limited to twelve. Time series are displayed on two charts where all data are shown in one chart, and a magnified subset is shown in the other chart (fig. 11). Borehole geophysical logs also can be viewed, cleaned, manipulated, and analyzed with SeriesSEE, where the two charts are displayed top-to-bottom, rather than left-to-right. SeriesSEE software, installation instructions, and help for all

tools can be downloaded in the zipped file, which is described in appendix A.

All source code that was developed for SeriesSEE can be downloaded freely (appendix B). All utilities, except WLM, are processed exclusively with VBA code in the SeriesSEE add-in or supporting add-in files named *SSmodule_*.SerSee*. Source codes for these files are in the VBA folder of appendix B and are named *SSmodule_*.xlsm*. Water levels to be modeled, input series, and period of analysis are defined with VBA routines. WLM components are transformed (table 1) and water levels are simulated with the FORTRAN program *WLmodel,* which reads ASCII files written by VBA programs. Differences between synthetic and measured water levels are minimized with *PEST* (Doherty, 2010a, 2010b). A copy of

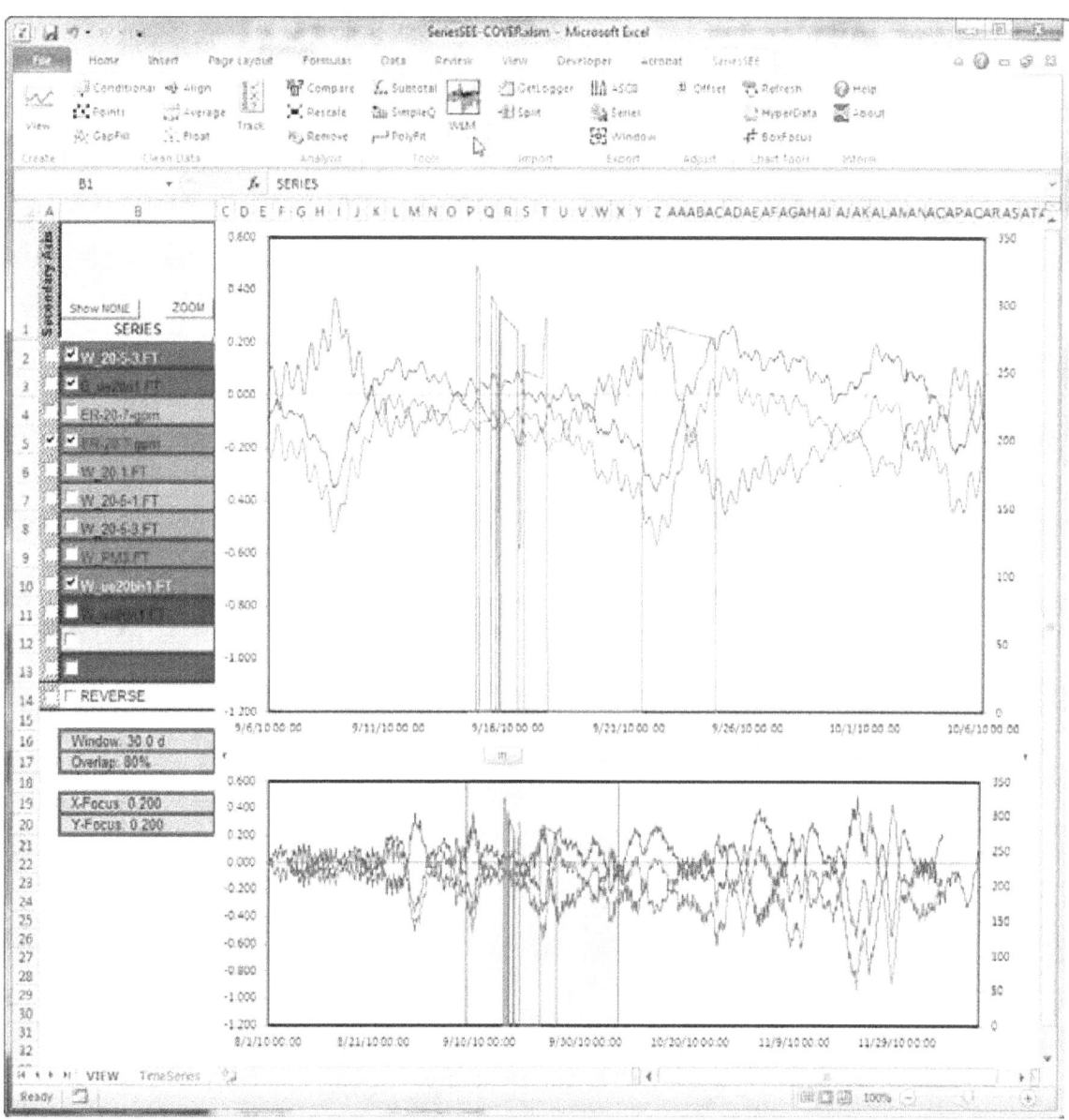

Figure 11.   SeriesSEE toolbar and example workbook that was created with SeriesSEE.

PEST exists in the SeriesSEE installation files, but also can be downloaded independently from *http://www.pesthomepage. org/*. The VBA utility WLM writes the PEST control file, **.pst*, as multiple, commented input files, which are concatenated and stripped of comments with the FORTRAN program *NoComment*. Source codes and documentation of *WLmodel* and *NoComment* are in the FORTRAN folder of appendix B.

Data Requirements

Data must be arranged as a continuous series of headers and values where all headers are in a single row (fig. 12). Multiple time columns can be specified, which allows for specification of series with different or irregular sampling intervals. All series are independent, so time columns need not be synchronous. Multiple data series can share a common time column (fig. 12, See columns C, D, and E), but the shared time column must be the first time column to the left of the data series.

A Viewer file is created by selecting a cell ⟨✎⟩ in the block of data to be analyzed and pressing the ⟨View⟩ button (fig. 11). The entire data block is copied from the user's original file into the viewer file by default. All equations within the block of data are converted to values in the viewer file, which breaks all linkages to the user's original workbook. Original data and formulas are not altered in the user's original file because all SeriesSEE operations act on a copy of the data in the viewer file.

Supporting Utilities

SeriesSEE features more than 20 supporting utilities in addition to the viewer creation and water-level modeling utilities already discussed (table 4). Many utilities exist to provide data-handling capabilities that can be used prior to water-level modeling. Related utilities are grouped and labeled as Clean Data, Analysis, Tools, Import, Export, Adjust, and Chart Tools (table 4).

Time-series data generally must be cleaned before analyzing. Cleaning removes erroneous measurements, converts units, reconciles continuous and periodic measurements, and removes step changes from transducer disturbances. All changes between the original and cleaned series can be recorded with explanations for each data change if the track utility is active. Changes and explanations are recorded to an auxiliary workbook that also contains the original and revised series. Utilities in the clean data and analysis groups perform these tasks (table 4).

Simple analysis and inspection of series are supported by utilities in the analysis group (table 4). New series can be created by adding, subtracting, multiplying, or dividing one series by another with the ⟨Compute⟩ utility. Measurement frequencies of the two series can differ because of interpolation. Smoother series can be created from noisy series with moving averages or LOWESS (LOcally Weighted Scatterplot Smoothing), which is a nonparametric method of fitting a curved line to data (Helsel and Hirsch, 1992, p. 288–291). Potential correlations among multiple series of disparate scales can be inspected by normalizing these series to a common scale with the ⟨Rescale⟩ utility.

Water-level modeling and other analyses can be expedited and improved by data reduction where there has been oversampling. Data can be reduced by averaging within periods such that 1-minute data are reduced to 1-hour averages with the ⟨Average⟩ utility. Continuous records of flow rates with many thousands of measurements can be reduced accurately to a few dozen step changes with the ⟨minimize⟩ utility. Simplified pumping schedules increase the efficiency and speed at which drawdowns can be simulated in WLMs. Geophysical logs are approximated with a simple polyline using the PolyFit utility, ⟨PolyFit⟩, which can eliminate extraneous fluctuations and constrain the polyline to monotonic increases. Utilities in the tools group perform these tasks (table 4).

Time series can be imported from ASCII files and database tables to a SeriesSEE data table with utilities in the import group (table 4). Multiple data-logger files are read interactively with the ⟨Import⟩ utility to create a single SeriesSEE data table. Database tables with site identifiers, times, and water levels grouped into three columns can be reformatted to a SeriesSEE data table with the ⟨Split⟩ utility.

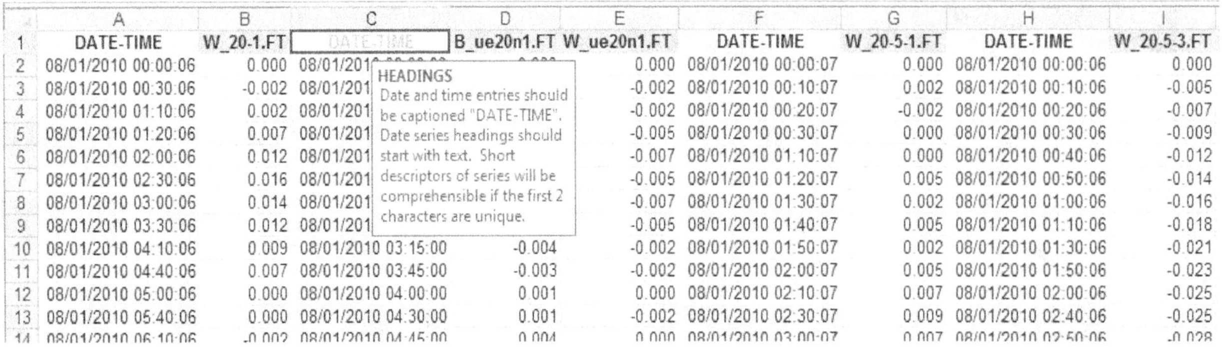

	A	B	C	D	E	F	G	H	I
1	DATE-TIME	W_20-1.FT	DATE-TIME	B_ue20n1.FT	W_ue20n1.FT	DATE-TIME	W_20-5-1.FT	DATE-TIME	W_20-5-3.FT
2	08/01/2010 00:00:06	0.000	08/01/2010 00:00:00	0.000	0.000	08/01/2010 00:00:07	0.000	08/01/2010 00:00:06	0.000
3	08/01/2010 00:30:06	-0.002	08/01/201		-0.002	08/01/2010 00:10:07	0.002	08/01/2010 00:10:06	-0.005
4	08/01/2010 01:10:06	0.002	08/01/201		-0.002	08/01/2010 00:20:07	-0.002	08/01/2010 00:20:06	-0.007
5	08/01/2010 01:20:06	0.007	08/01/201		-0.005	08/01/2010 00:30:07	0.000	08/01/2010 00:30:06	-0.009
6	08/01/2010 02:00:06	0.012	08/01/201		-0.007	08/01/2010 01:10:07	0.000	08/01/2010 00:40:06	-0.012
7	08/01/2010 02:30:06	0.016	08/01/201		-0.005	08/01/2010 01:20:07	0.005	08/01/2010 00:50:06	-0.014
8	08/01/2010 03:00:06	0.014	08/01/201		-0.007	08/01/2010 01:30:07	0.002	08/01/2010 01:00:06	-0.016
9	08/01/2010 03:30:06	0.012	08/01/201		-0.005	08/01/2010 01:40:07	0.005	08/01/2010 01:10:06	-0.018
10	08/01/2010 04:10:06	0.009	08/01/2010 03:15:00	-0.004	-0.002	08/01/2010 01:50:07	0.002	08/01/2010 01:30:06	-0.021
11	08/01/2010 04:40:06	0.007	08/01/2010 03:45:00	-0.003	-0.002	08/01/2010 02:00:07	0.005	08/01/2010 01:50:06	-0.023
12	08/01/2010 05:00:06	0.000	08/01/2010 04:00:00	0.001	0.000	08/01/2010 02:10:07	0.007	08/01/2010 02:00:06	-0.025
13	08/01/2010 05:40:06	0.000	08/01/2010 04:30:00	0.001	-0.002	08/01/2010 02:30:07	0.009	08/01/2010 02:40:06	-0.025
14	08/01/2010 06:10:06	-0.002	08/01/2010 04:45:00	0.004	0.000	08/01/2010 03:00:07	0.007	08/01/2010 02:50:06	-0.028

HEADINGS
Date and time entries should be captioned "DATE-TIME". Date series headings should start with text. Short descriptors of series will be comprehensible if the first 2 characters are unique.

Figure 12. Format of headers and values for creating a viewer file with SeriesSEE.

Table 4. Summary of available tools in SeriesSEE.

Group	Utility	Description	Name
Create		Create Viewer file by selecting a cell in a block of data in an original source file, which is copied to the viewer file. All equations are converted to values in the Viewer file.	View
Clean Data		Bad data conditionally can be commented and/or eliminated.	Conditional
		Bad data in a single series can be commented and/or eliminated graphically.	Points
		Data gaps from the cleaning process can be filled by linear interpolation, loaded with a dummy value, eliminated altogether, or gaps can be created for alignment.	GapFill
		Shift data segments. Estimate shift with simple water-level models that use a few guide series. Alternatively, shifts can be assigned from other estimates.	Align
		Data reduction by averaging where oversampled.	Average
	Float	Float series to tape downs without changing slope of transducer data.	Float
	Track	Force an explanation to be appended to each data change in an auxillary workbook that also contains the original and revised series.	Track
Analysis		Create new series by addition, subtraction, multiplication, and division of existing series. Second series interpolated to times in the first series. Series can also be smoothed with a moving average or LOWESS curve.	Compare
		Series can be normalized to common scales.	Rescale
		Removes derived series that are created by Compare or Rescale.	Remove
Tools		Data reduction tool where selected series are binned by time periods or depth intervals to compute statistics.	Subtotal
		Reduces pumping rates to a simple schedule.	SimpleQ
		Geophysical logs are approximated with a simple polyline.	PolyFit
	WLM	Model water levels interactively in a new workbook, where water levels are simulated with a FORTRAN program and differences are minimized with PEST.	WLM
Import		Series from data-logger files are read interactively and concatenated in a SeriesSEE format.	GetLogger
	Split	Split 3 columns of site identifiers, time, and water levels into SeriesSEE input where a new series is identified at each change in site identifier.	Split
Export	ASCII	Output from tracking workbooks to selected ASCII formats.	ASCII
		Export individual series with options to create drawdown observations. Drawdown observations require shifting, binning, and truncating to a time window.	Series
		Data are copied to a new workbook and reduced to a user-specified period.	Window
Adjust		Individual, selected, or all series can be shifted such that the average, minimum, maximum, or first value will equal zero.	Offset
Chart Tools		Refresh the list of available series after manually adding or deleting series on the data page.	Refresh
		Create temporary hyperlinks between visible series and charted data in the Viewer file.	HyperData
		Magnify subareas of plot. First click adds a rectangle. Second click re-scales both axes to rectangle area. Third click restores plot.	BoxFocus
Inform	Help	Controls and usage of SeriesSEE are explained.	Help
		Display ad copy about SeriesSEE.	About

Series can be viewed and inspected at scales as fine as discrete measurements with utilities in the "adjust" and "chart tools" groups (table 4). Series can be shifted so that all measurements fluctuate about a common reference with the ⊞ Offset utility, which eases comparisons among series (fig. 13). Subareas of charts in SeriesSEE viewer and auxiliary files can be magnified interactively with the ⌐⌐ BoxFocus utility. Discrete measurements can be selected graphically and connected to the cell with the numerical value in the Viewer file with the ◰ HyperData utility, which creates temporary hyperlinks between charted points and the cell with the plotted value.

Each SeriesSEE utility is fully documented in the help system, which can be called with the ⊘ Help utility or from context sensitive help calls in each utility (appendix A). Each group, utility, form, and auxiliary workbook is explained briefly, and step-by-step instructions (fig. 14). Complex utilities such as water-level modeling are documented with multiple pages that explain each form and action.

Water-Level Modeling

Water levels are modeled interactively with the WLM utility in SeriesSEE. Water levels to be modeled, input series, period of analysis, and WLM components are defined through the use of data-entry forms. A new workbook for modeling water levels is created with user-specified information from these forms. Fitting periods and WLM components can be modified in the WLM workbook.

Analytical models that transform WLM components in the FORTRAN program *WLmodel* have been verified (table 5). The analytical models for moving average and step transforms were verified against intrinsic functions in Excel. The analytical models for Theis, tide, pneumatic lag, and gamma transforms were verified against solutions that were computed with published programs. Source problems, programs, and comparisons between *WLmodel* output and published programs are documented fully in appendix C.

Differences between synthetic and measured water levels are minimized with PEST. Parameter estimates, transformed WLM components, synthetic water levels, and differences are imported automatically into the WLM workbook after PEST finishes. Model fit is defined by RMS error and evaluated graphically. Parameters are estimated and WLM results are evaluated iteratively until the user deems the fit to be adequate.

Table 5.   Summary of verification tests for analytical models in the FORTRAN program WLmodel.

WLM Component	SeriesSEE Label	Verification Source
Moving Average	SERIES	Excel function
Theis Transform	THEIS	Barlow and Moench, 1999
Tide	TIDE	Harrison, 1971
Step	STEP	Excel function
Pneumatic Lag	AIR-LAG	Barlow and Moench, 1998
Gamma	GAMMA	O'Reilly, 2004

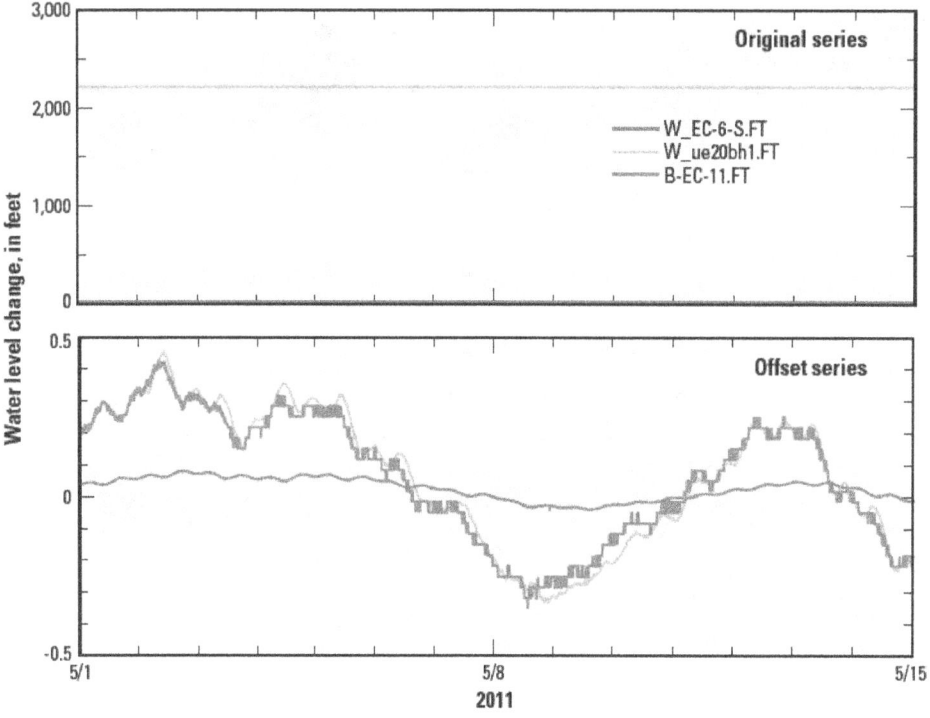

Figure 13.   Shifting series to a common reference with the offset utility.

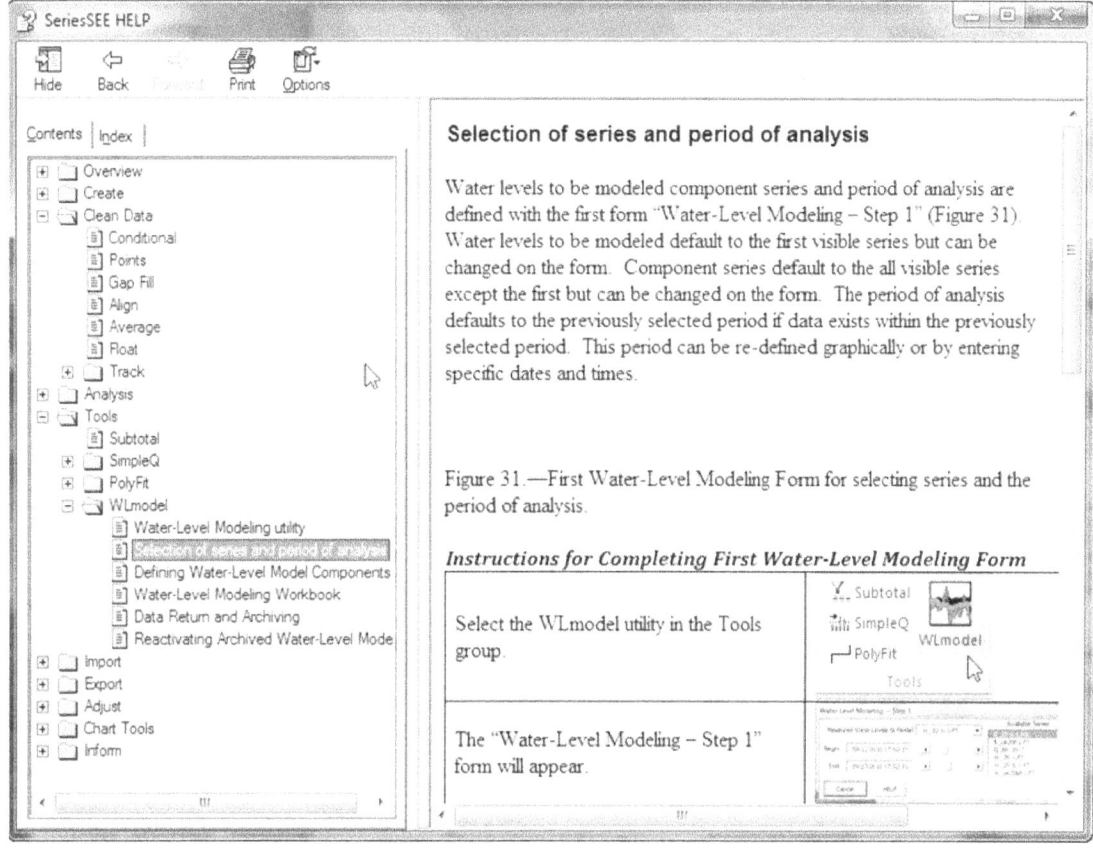

Figure 14. Table of contents and an explanation page in the help system for SeriesSEE.

Drawdowns and transformed WLM components are returned to the SeriesSEE viewer once the user accepts a WLM, where drawdowns are the sum of all Theis transforms minus differences between synthetic and measured water levels. Drawdowns and transformed WLM components are selected individually, so the number of returned series can range from 0 to all WLM components. The WLM workbook can be archived as a macro-free workbook with re-activation capabilities.

Applications of Water-Level Modeling

Water-level modeling applications of SeriesSEE are demonstrated with a hypothetical example and a field investigation at Pahute Mesa, Nevada National Security Site (NNSS). The hypothetical example emulated the complex hydrogeology beneath Pahute Mesa so that known drawdowns could be simulated in a complex aquifer system. Limitations of the Theis-transform approach were investigated with these known drawdowns. Environmental noise, which was the record of water levels in background well ER EC-6 shallow (table 6), was added to known drawdowns. The field investigation demonstrated that drawdowns much smaller than environmental

fluctuations can be detected across a major fault structure more than 1 mile from the pumping well. Explanations, data sets, and ancillary software for the hypothetical example and field investigation are in appendixes D and E, respectively.

Water-level modeling was developed and tested with data from Pahute Mesa, NNSS, (fig. 15) because detection of distant drawdowns is imperative and complicated by more than 2,000 ft of unsaturated zone. Migration of radionuclides from underground testing of nuclear devices drives the need to quantify groundwater flow and transport beneath Pahute Mesa (Laczniak and others, 1996). The great depth to water and accessibility limit the number of wells, which typically penetrate a mile of volcanic rock and are more than 1-mi apart (Fenelon and others, 2010). Environmental water-level fluctuations are substantial beneath Pahute Mesa because of the thick unsaturated zone and high hydraulic diffusivity of the volcanic rocks.

The aquifer system beneath Pahute Mesa comprises layered sequences of volcanic rocks that have been faulted into distinct structural blocks (Warren and others, 2000). Rhyolitic lavas or welded ash-flow tuffs such as in the Benham and Topopah Springs aquifers, respectively, comprise aquifers. Bedded and non-welded, zeolitized tuffs typically comprise confining units (Blankennagel and Weir, 1973; Prothro and Drellack, 1997; Bechtel Nevada, 2002). More than a half

Table 6. Site information and completion depths for wells at Pahute Mesa, Nevada National Security Site that were used in hypothetical example and field investigation.

Well name: Names are listed in alphabetical order Bold part of name is well site as shown on Figure 15
U S Geological Survey site identification number: Unique 15-digit number identifying well
Latitude/Longitude: Latitude and longitude coordinates, referenced to North American Datum of 1927
Land-surface altitude: Altitude, referenced to National Geodetic Vertical Datum of 1929
Open intervals: Depth, in feet below land surface, of the top and bottom of open annulus

Well Name	U.S. Geological Survey site identification number	Latitude (degrees, minutes, seconds)	Longitude (degrees, minutes, seconds)	Land-surface altitude (feet)	Open intervals
ER-20-5 #1	371312116283801	37 13'12.2"	116 28'37.8"	6,242	2,249–2,655
ER-20-6 #3	371533116251801	37 15'33.1"	116 25'17 5"	6,466	2,436–2,807
ER-EC-6 shallow	371120116294805	37 11'19.6"	116 29'48 1"	5,604	1,606–1,948
ER-EC-11 main	371151116294102	37 11'51.2"	116 29'41 1"	5,656	3,196–3,385 3,590–4,148
PM-3-1	371421116333703	37 14'20.7"	116 33'36.6"	5,823	1,872–2,192
UE-20n 1	371425116251902	37 14'25.1"	116 25'19.0"	6,461	2,308–2,834
ER-20-7	371247116284502	37 12'47.0"	116 28'44.8"	6,209	2,292–2,924
ER-20-8 main	371135116282601	37 11'35.1"	116 28'26 3"	5,848	2,440–2,940 3,070–3,442

dozen faults with offsets in excess of 500 ft have been mapped previously in Pahute Mesa (McKee and others, 2001), and additional faults are mapped with each new well (for example, National Security Technologies, LLC, 2010).

Hypothetical Example

The reliability of differentiating environmental fluctuations and pumping responses with water-level models was tested with a hypothetical aquifer system. Drawdown from a hypothetical aquifer test was simulated where the hydrogeologic complexity and distribution of hydraulic properties were assigned. The hypothetical aquifer system is comprised of ash-fall tuff, bedded tuff, welded tuff, and lava units that are flat-lying, laterally isotropic, and homogeneous (fig. 16). A fault 1,500 ft east of the pumping well, P1, bisects the aquifer system, vertically displaces hydrogeologic units 1,000 ft, and alters hydraulic properties around the structure.

The hypothetical aquifer system was simulated with a three-dimensional MODFLOW model (Harbaugh, 2005). The model domain was discretized laterally into 135 columns of 135 rows with a variably spaced grid (fig. 16). Cell sizes ranged in width from 10 ft by the pumping well to 40,000 ft at the model edges. Model edges were about 200,000 ft away from the pumping well, P1, and were simulated as no-flow boundaries. The model grid extended vertically from an impervious base at sea level to the water table at 4,200 ft above sea level. Vertical discretization was uniform, with 200-ft thick layers except for a 1-ft thick layer at the water table. The thickness differed so that the storage coefficient and specific storage were equivalent, and it allowed specific yield to be assigned directly in a layer. Changes in saturated thickness of the aquifer were not simulated because maximum drawdown at the water table was small relative to the total thickness.

Hydraulic properties typical of volcanic units were assigned to the hypothetical aquifer system. Ash-fall tuff, bedded tuff, welded tuff, and lava were assigned hydraulic conductivities of 0.001, 0.1, 3, and 50 ft/d, respectively. Horizontal-to-vertical anisotropy of one was assigned to all units. A uniform value of 0.02 was assigned for specific yield. The specific storage of all hydrogeologic units was 2×10^{-6} 1/ft.

Hypothetical aquifer-test results were simulated and analyzed during a 3-month period that was divided into five stress periods. The antecedent, pumping, recovery, pumping, and recovery periods were 21, 10, 10, 10, and 40 days, respectively. Pumping rates were 500 gpm during the second and fourth stress periods. Flow and drawdown in pumping and observation wells were simulated and sampled with the Multi-Node Well (MNW) package (Harbaugh, 2005). Flow to the pumping well was distributed proportionally to cell transmissivities by the MNW package.

Water levels with a "known" pumping signal and environmental fluctuations (noise) shown in figure 17 for well O3 were created by adding simulated drawdowns from MOD-FLOW to measured water levels in well EREC-6 shallow (fig. 17). Simulated drawdowns from MODFLOW in well O3, which is 7,800 ft from well P1, were interpolated in time to match measured water levels in well EREC-6 shallow. Simulated drawdowns from MODFLOW and simulated drawdowns with environmental noise added are in appendix D in the file .\ WLMs\00_Hypo+Meas2SeriesSEE.xlsx.

Drawdowns were estimated by modeling "measured" water levels in well O3. Environmental fluctuations were simulated with computed tides, barometric pressure and background water levels in wells PM-3 and UE-20n 1 (fig. 17). Pumping effects were simulated with a Theis transform of the hypothetical pumping schedule. The water-level model was calibrated during the period from November 18, 2010, to March 6, 2011.

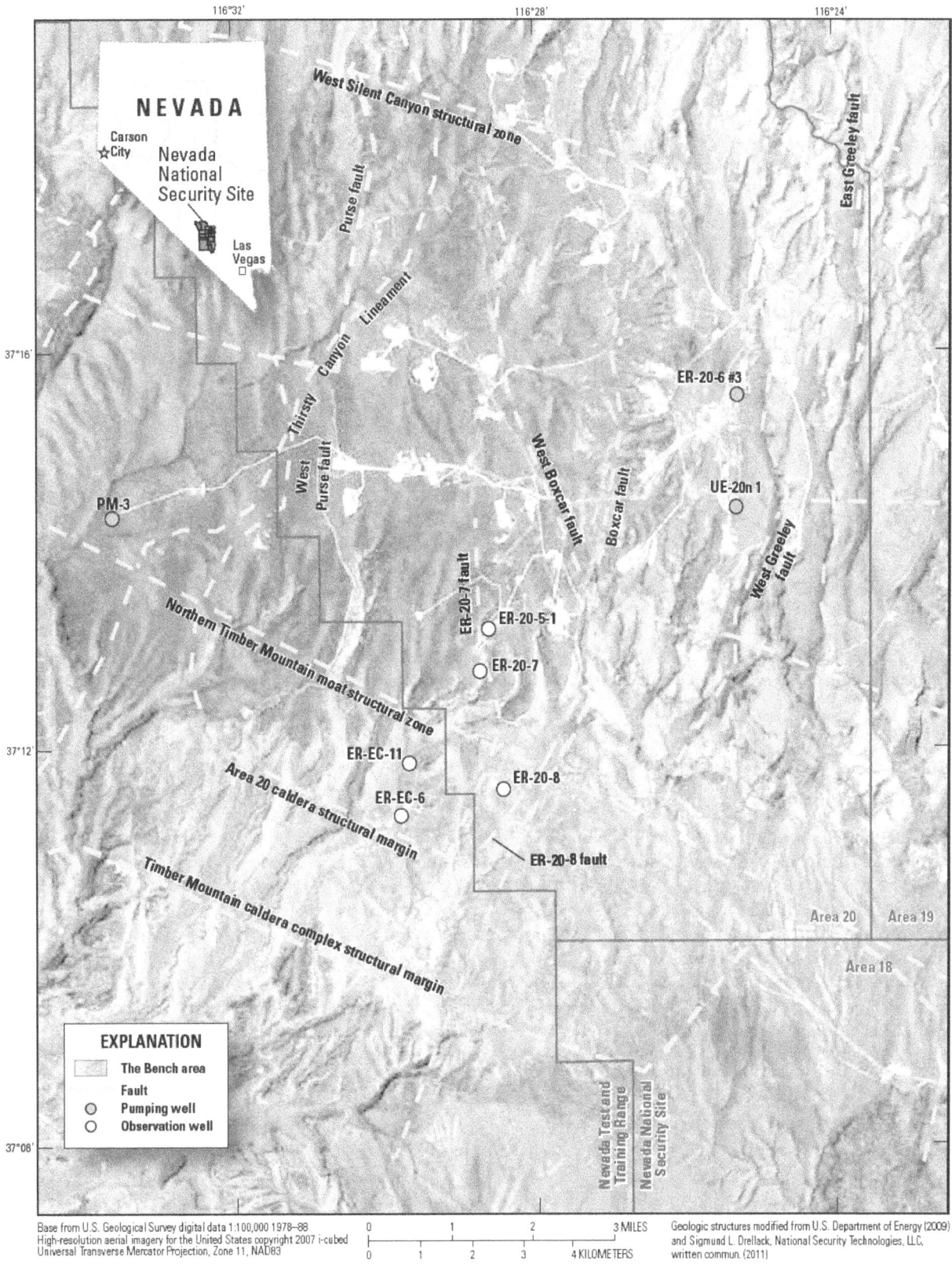

Figure 15. Background wells, observation wells, pumping well, and selected fault structures at Pahute Mesa Nevada National Security Site.

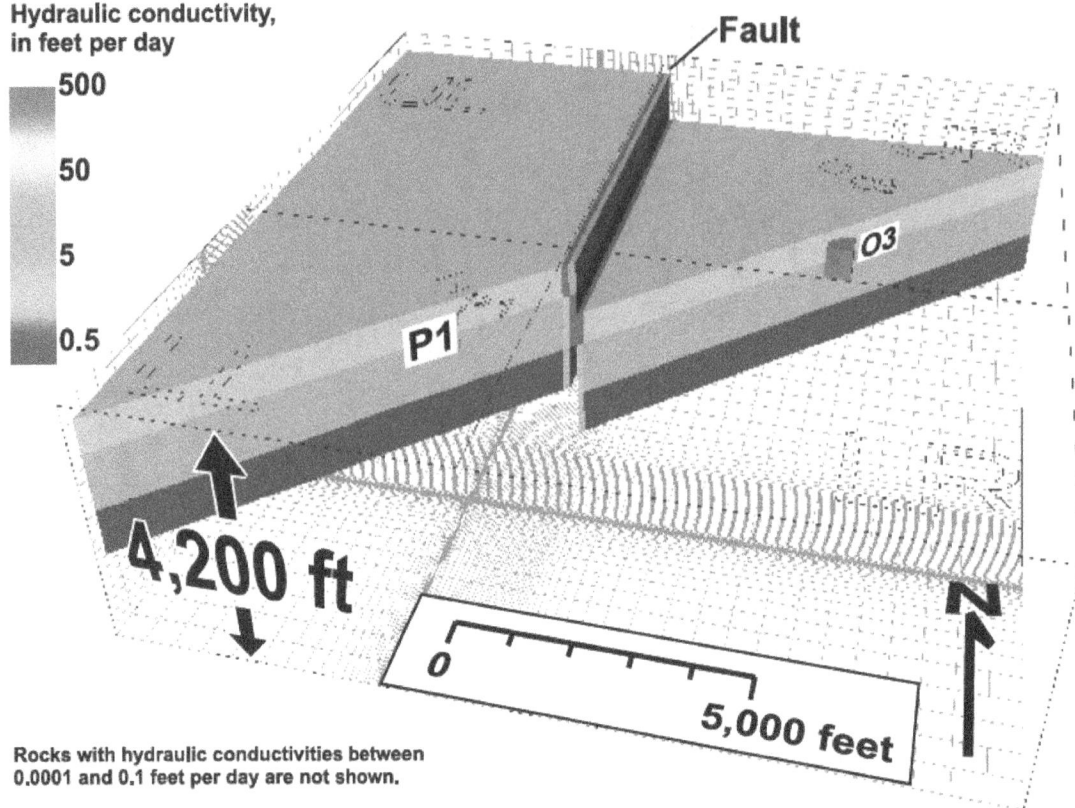

Figure 16.   Hydraulic conductivity distribution of a subset of a hypothetical aquifer system that has been bisected by a fault, showing well locations and labeled quadrants (upper left, UL; upper right, UR; lower right, LR; lower left, LL).

Drawdowns that were estimated from "measured" water levels in well O3 agreed with known drawdowns within the noise of the data set (fig. 18). A maximum drawdown of 0.18 ft was estimated which was identical to the known maximum. The RMS error of differences between synthetic and measured water levels was 0.013 ft. The RMS error of differences between synthetic and known drawdowns was 0.015 ft.

Drawdowns alternatively were estimated in well O3 by modeling the original MODFLOW results with Theis transforms. No other WLM components were considered because environmental fluctuations did not exist in the original MODFLOW results. This alternative water-level model also was calibrated during the period from November 18, 2010, to March 6, 2011.

Drawdowns that were estimated directly from MODFLOW results could be replicated almost perfectly with Theis transforms. Differences between MODFLOW results and a single Theis transform could be reduced to a RMS error of less than 0.006 ft. RMS error declined to less than 0.0006 ft with the addition of a second Theis transform (fig. 18). Deviations of less than 0.001 ft approach the accuracy of the numerical solution of the hypothetical aquifer test.

The simplicity of Theis transforms did not introduce error because MODFLOW results could be replicated near perfectly with Theis transforms. Differences between known drawdowns and drawdowns that were estimated from "measured" water levels differed because of noise in the measured input series.

The hypothetical model and SeriesSEE input were created with **HypoFrame**, which is a workbook for simulating hypothetical aquifer tests and creating water levels with known pumping signals and environmental noise. Hypothetical aquifer systems must have flat-lying geologic units of uniform thickness and laterally isotropic, homogeneous hydraulic conductivity. A hypothetical aquifer system can be subdivided into four quadrants by two intersecting faults. Rock sequences in each quadrant can be displaced vertically within each quadrant. The **HypoFrame** workbook and documentation are in appendix D.

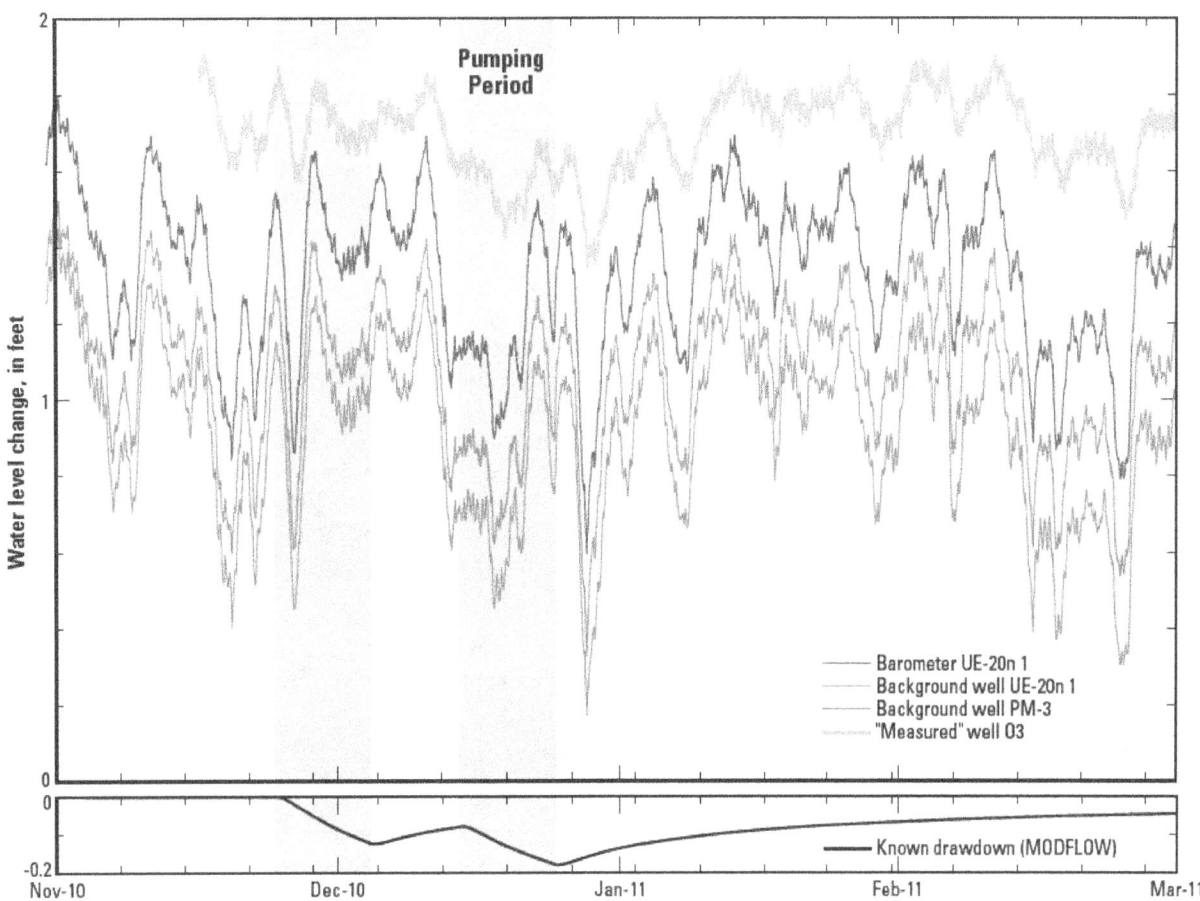

Figure 17. Barometric pressure, background water levels, and water levels with known drawdowns in hypothetical well O3.

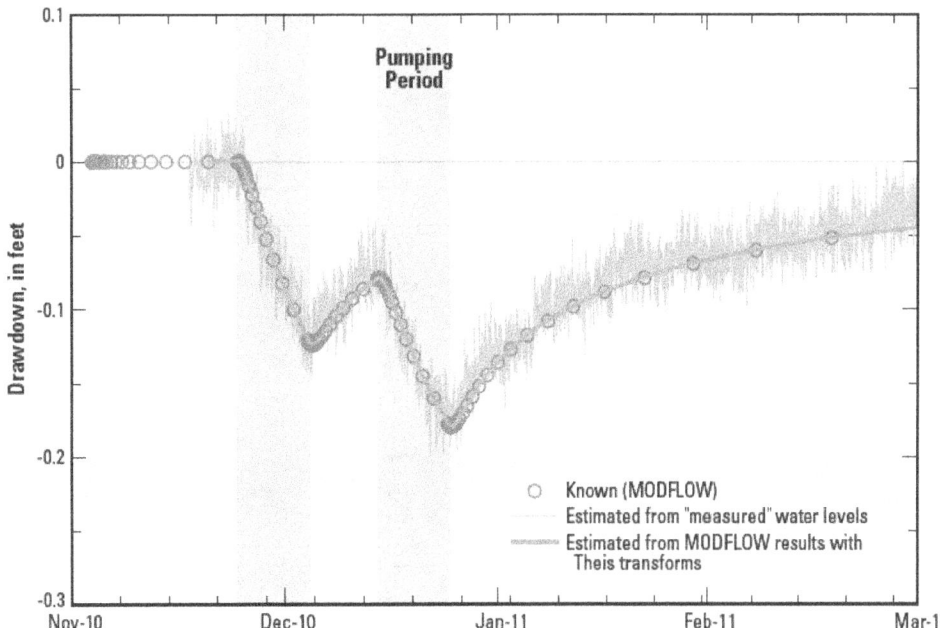

Figure 18. Known drawdowns (MODFLOW), drawdowns estimated from "measured" water levels, and drawdowns estimated directly from MODFLOW results in well O3.

Pahute Mesa Example

Water-level modeling was tested in a complex hydrogeologic system by estimating drawdown from two aquifer tests beneath Pahute Mesa (Halford and others, 2011). The upper and lower zones of well ER-20-8 main produced water from the Tiva Canyon and Topopah Spring aquifers sequentially between June 16, 2011, and August 8, 2011. Each well was pumped a total of 20 d, where pumping periods were evenly divided between well development and a constant-rate test (fig. 19).Drawdown from pumping both zones was estimated in observation well ER-20-7, which is screened in the Topopah Spring aquifer. Pumping and observation wells are 1.4 mi apart and penetrate different structural blocks (fig. 15).

Drawdown in well ER-20-7 was estimated with multiple Theis transforms in the water-level model. Environmental fluctuations were simulated with computed tides, barometric pressure, and background water levels from well UE-20bh-1 (fig. 15). Pumping effects were simulated with two Theis

transforms for each of the two pumping schedules (fig. 19). The fitting period was from April 20, 2011, to November 11, 2011. Synthetic water levels matched measured water levels with a RMS error of 0.004 ft.

Drawdown in well ER-20-7 also was estimated with an identical water-level model, except that WLM components with background water levels were negated. Synthetic water levels matched measured water levels with a RMS error of 0.027 ft during the same fitting period from April 20, 2011, to November 11, 2011 (fig. 19). Each drawdown estimate was the difference between a synthetic water level without Theis transforms and a measured water level.

Poor drawdown estimates from the water-level model without background water levels demonstrates the need to simulate as much of the environmental fluctuations as possible. Antecedent conditions were simulated poorly where estimated drawdowns should be zero. Estimated drawdowns unambiguously were wrong during October and November when net water-level rises from pumping were estimated (fig. 19).

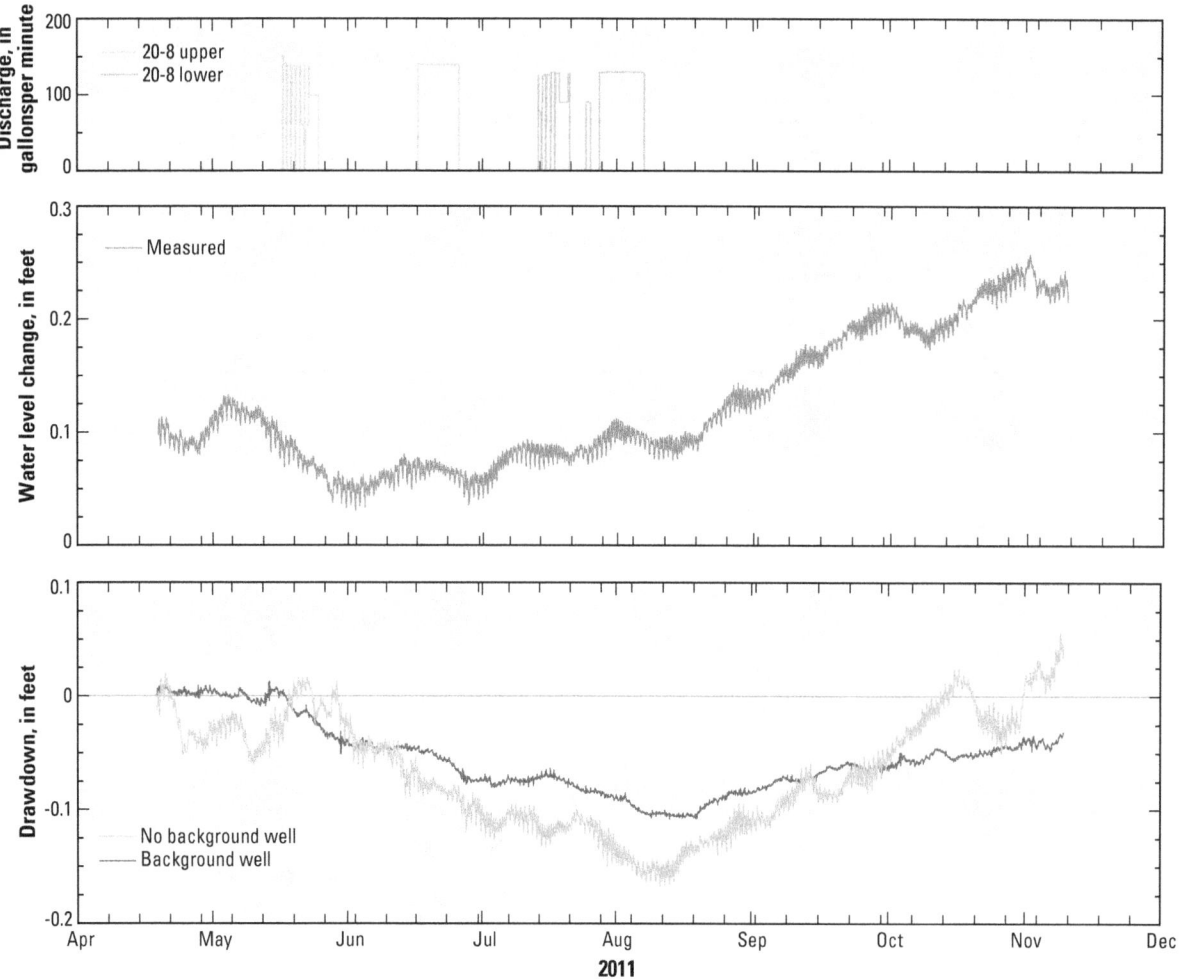

Figure 19.   Measured water levels, synthetic water levels, Theis transforms, and estimated drawdowns in well ER-20-7 from pumping ER-20-8 main upper and lower zones, Pahute Mesa, Nevada National Security Site.

Water-Level Modeling Strategies

Estimating drawdowns that have been obscured by environmental fluctuations is the primary goal of the water-level modeling approach. This approach is most effective and efficient where many WLM components are specified and fitting periods are great. This approach has been summarized, sometimes derisively, as the flak-gun, fishing-with-dynamite, and kitchen-sink approaches. All phrases accurately depict testing many WLM components simultaneously. Unique contributions from each WLM component remain unknown, but pumping signals are not correlated with environmental fluctuations. The flak-gun approach was adopted here.

The flak-gun approach uses WLM components that could have been excluded. This is not a problem because mechanisms exist to negate WLM components. Amplitudes tending to zero will negate a WLM component. Multiple WLM components also can negate one another by summing to zero. Likewise, Theis transforms also are negated by a large transmissivity or storage coefficient value where pumping signals are below detection or absent. Negated WLM components aesthetically are lacking, but do not affect results. Systematic investigation of WLM components is possible with SeriesSEE, but has not been automated.

The flak-gun approach has many advantages, especially when estimating drawdowns in dozens of wells. Reporting is easier because the same input series and WLM components were used in all of the water-level models. Water-level models calibrate quickly after analyzing the first or second well because WLM components are defined with fair initial estimates of amplitude and phase. The flak-gun approach can fail when the fitting period decreases and correlation becomes possible between pumping signals and environmental fluctuations.

Correlation between weak pumping signals and environmental fluctuations is possible and requires further investigation. Nebulous drawdown estimates can be investigated with multiple water-level models where water levels initially are simulated without Theis transforms. An alternative water-level model is created by adding a Theis transform to the initial water-level model. The initial transmissivity and storage coefficient should create a small but measureable maximum deflection in the added Theis transform. Drawdowns likely were not detected if the RMS error cannot be reduced by more than 30 percent.

Input series of greater duration potentially can degrade with time as pressure transducers fail. For example, multiple input series could be good for the first four months, while one input series degrades during the last two months. Degradation likely will be apparent in the WLM residuals as scatter increases. Identifying the onset of failure in a specific input series requires modeling water levels during subsets of the fitting period. Degrading input series can be investigated manually with SeriesSEE, but an automated tool would be a better approach.

Summary and Conclusions

Pumping responses can be differentiated reliably from environmental fluctuations with water-level modeling. Water-level modeling approximates measured water-level fluctuations by summing multiple component fluctuations. Environmental fluctuations primarily are composed of barometric and background water-level input series and computed tide components. Pumping signals are modeled by superimposing multiple Theis transforms, where step-wise pumping records of flow are transformed into water-level changes. The summation of all component fluctuations is a synthetic water-level series.

Water-levels can be modeled robustly with the Theis-transform approach because environmental fluctuations and pumping signals are simulated simultaneously. Long-term trends are well simulated because environmental fluctuations are defined with entire periods of record. Fitting periods are extended greatly where pumping and recovery affect a majority of the record. Multiple Theis responses with different hydraulic diffusivities are summed to approximate lithologic variability.

Water-level modeling with Theis transforms has been implemented in the program SeriesSEE, which is a Microsoft® Excel add-in. Water levels to be modeled, input series, period of analysis, and water-level model components are defined interactively and viewed in workbooks that are created by SeriesSEE. Water levels are modeled with a FORTRAN program that is called from Excel. Differences between synthetic and measured water levels are minimized with PEST.

Water-level model components are transformations of input series. Moving average, Theis, pneumatic-lag, and gamma transforms are available transforms in SeriesSEE. Moving averages most frequently transform input series of barometric pressure and background water levels. Pumping schedules are transformed into water-level fluctuations with Theis transforms. Pneumatic-lag transforms barometric pressure changes at land surface to lagged and attenuated responses at the water table. Water-level rises from infiltration events are simulated with the gamma transform. Earth tides and step transforms are purely computed quantities that do not require input series.

Many utilities exist in SeriesSEE for viewing, cleaning, manipulating, and analyzing time-series data in addition to water-level modeling. Supporting utilities exist because data handling frequently consumes more time and effort than water-level modeling. Each SeriesSEE utility is documented with a brief explanation and step-by-step instructions that are accessed through context sensitive help.

Water-level models must be calibrated to reliably differentiate small pumping responses from environmental fluctuations. Differences between synthetic and measured water levels define goodness-of-fit. Sum-of squares of differences are minimized by PEST where singular value decomposition and Tikhonov regularization are used to assure stable results, not to inform estimated parameter values. Preferred homogeneity within amplitude, phase, and hydraulic property parameters is enforced with Tikhonov regularization.

Drawdown estimates from a water-level model are the summation of all Theis transforms minus residuals. The summation of all Theis transforms is the direct estimate of the pumping signal. Residuals represent all unexplained water-level fluctuations. These fluctuations should be random residuals during non-pumping periods, but can contain unexplained components of the pumping signal during pumping and recovery periods.

The simplicity of Theis transforms did not introduce error because results from a hydrogeologically complex MODFLOW model could be replicated near perfectly with Theis transforms. Differences between known drawdowns and drawdowns that were estimated from "measured" water levels differed because of noise in the measured input series. Estimated drawdowns are affected minimally by the Theis-transform approach relative to the inaccuracies that result from noise in the data sets.

Drawdowns much smaller than environmental fluctuations have been detected across a major fault structure more than 1 mile from the pumping well beneath Pahute Mesa, Nevada National Security Site. A maximum drawdown of 0.1 ft was estimated in well ER-20-7 during an 8-month period of analysis. Drawdown estimates in well ER-20-7 were consistent with a plausible pattern of drawdowns at all observation wells. Drawdowns could not have been detected without water-level modeling as implemented in SeriesSEE.

References

Baehr, A.L., and Hult, M.F., 1991, Evaluation of unsaturated zone air permeability through pneumatic tests: Water Resources Research, v. 27, no. 10, p. 2605–2617.

Barlow, P.M., and Moench, A.F., 1998, Analytical Solutions and Computer Programs for Hydraulic Interaction of Stream-Aquifer Systems: United States Geological Survey, Open-File Report 98-415A, 85 p.

Barlow, P.M., and Moench, A.F., 1999, WTAQ—A computer program for calculating drawdowns and estimating hydraulic properties for confined and water-table aquifers: U.S. Geological Survey Water-Resources Investigations Report 99-4225, 84 p.

Bartels, J., 1957, Gezeitenkrafte: Encyclopedia of Physics, vol. 48, 734: Berlin, Springer.

Bartels, J., 1985, Tidal forces (English translation), in Harrison J.C. (ed.) Earth Tides: New York, Van Nostrand Reinhold, p. 25–63.

Beaumont, C., and Berger, J., 1975, An analysis of tidal strain observations from the United States of America: I. The laterally homogeneous tide: Bulletin of the Seismological Society of America, v. 65, no. 6, p. 1613–1629.

Berger, J., and Beaumont, C., 1976, An analysis of tidal strain observations from the United States of America II. The inhomogeneous tide: Bulletin of the Seismological Society of America, v. 66, no. 6, p. 1821–1846.

Bechtel Nevada, 2002, A hydrostratigraphic model and alternatives for the groundwater flow and contaminant transport model of Corrective Action Units 101 and 102—Central and western Pahute Mesa, Nye County, Nevada: U.S. Department of Energy Report DOE/NV/11718–706.

Blankennagel, R.K., and Weir, J.E., Jr., 1973, Geohydrology of the eastern part of Pahute Mesa, Nevada Test Site, Nye County, NV: U.S. Geological Survey Professional Paper 712-B (Also available at *http://pubs.usgs.gov/pp/0712b/ report.pdf.*)

Bower, D.R., 1983, Bedrock fracture parameters from the interpretation of well tides: Journal of Geophysical Research, v. 88, no. B6, p. 5025–5035.

Bohling, G.C., Zhan, X., Knoll, M.D., and Butler J.J., 2003, Hydraulic tomography and the impact of a priori information: An alluvial aquifer example. Kansas State Geological Survey Open-File Report 2003-71. Lawrence, Kansas: Kansas State Geological Survey.

Bredehoeft, J.D., 1967, Response of well-aquifer systems to earth tides: Journal of Geophysical Research, v. 72, no. 12, p. 3075–3087.

Criss, R. E. and Criss, E.M., 2011. Prediction of well wevels in the alluvial aquifer along the Lower Missouri River: Ground Water. doi: 10.1111/j.1745-6584.2011.00877.x

Defant, A., 1958, Ebb and Flow: Ann Arbor, University of Michigan Press.

Doherty, J., 2010b, Addendum to the PEST manual: Brisbane, Australia, Watermark Numerical Computing.

Doherty, J., 2010a, PEST, Model-independent parameter estimation—User manual (5th ed., with slight additions): Brisbane, Australia, Watermark Numerical Computing.

Doherty, J.E., and Hunt, R.J., 2010, Approaches to highly parameterized inversion—A guide to using PEST for groundwater-model calibration: U.S. Geological Survey Scientific Investigations Report 2010–5169, 59 p.

Doherty, J. and Johnston, J.M., 2003, Methodologies for calibration and predictive analysis of a watershed model: Journal of the American Water Resources Association, v. 39, no. 2, p. 251–265.

Dooge, J.C.I., 1959, A general theory of the unit hydrograph: Journal of Geophysical Research, v. 64, no. 2, p. 241–256.

Elliot, P.E., and Fenelon, J.M., 2010, Database of groundwater levels and hydrograph descriptions for the Nevada Test Site area, Nye County, Nevada, 1941–2010: U.S. Geological Survey Data Series 533. (Also available at *http://pubs.usgs.gov/ds/533/*.)

Erskine, A. D., 1991, The effect of tidal fluctuation on a coastal aquifer in the UK: Ground Water, v. 29, p. 556–562. doi: 10.1111/j.1745-6584.1991.tb00547.x

Fenelon, J.M., 2000, Quality assurance and analysis of water levels in wells on Pahute Mesa and vicinity, Nevada Test Site, Nye County, Nevada: U.S. Geological Survey Water-Resources Investigations Report 00-4014. (Also available at *http://pubs.usgs.gov/wri/WRIR00-4014/*.)

Fenelon, J.M., 2005, Analysis of ground-water levels and associated trends in Yucca Flat, Nevada Test Site, Nye County, Nevada, 1951–2003: U.S. Geological Survey Scientific Investigations Report 2005-5175, 87 p., at URL: *http://pubs.water.usgs.gov/sir2005-5175*.

Fenelon, J.M., Sweetkind, D.S., and Laczniak, R.J., 2010, Groundwater flow systems at the Nevada Test Site, Nevada: A synthesis of potentiometric contours, hydrostratigraphy, and geologic structures: U.S. Geological Survey Professional Paper 1771.

Halford, K.J., 2006, Documentation of a spreadsheet for time-series analysis and drawdown estimation: U.S. Geological Survey Scientific Investigations Report 2006-5024. (Also available at http://pubs.usgs.gov/sir/2006/5024/PDF/SIR2006-5024.pdf.)

Halford, K.J., Fenelon, J.M., and Reiner, S.R., 2010, Aquifer-test package—Analysis of ER-20-8 #2 and ER-EC-11 multi-well aquifer tests, Pahute Mesa, Nevada National Security Site: unpublished U.S. Geological Survey aquifer-test package, accessed August 30, 2011, at URL: *http://nevada.usgs.gov/water/AquiferTests/er_wells.cfm?studyname=er_wells*.

Halford, K.J., and Yobbi, D.K., 2006, Estimating hydraulic properties using a moving-model approach and multiple aquifer tests: Ground Water, v. 44, no. 2, p. 284–291.

Hanson, J.M., and Owen, L.B., 1982, Fracture orientation analysis by the solid earth tidal strain method: Presented at the 57th Annual Fall Technical Conference and Exhibition of the Society of Petroleum Engineers of AIME, American Institute of Mechanical Engineers, New Orleans, Louisiana, September 26–29, 1982.

Harbaugh, A.W., 2005, MODFLOW-2005, the U.S. Geological Survey modular ground-water model -- the ground-water flow process: U.S. Geological Survey Techniques and Methods 6-A16.

Harp, D. R., and Vesselinov, V.V., 2011, Identification of pumping influences in long-term water level fluctuations: Ground Water, v. 49, p. 403–414. doi: 10.1111/j.1745-6584.2010.00725.x

Harrison, J.C., 1971, New computer programs for the calculation of earth tides: Cooperative Institute for Research in Environmental Sciences, National Oceanic and Atmospheric Administration/University of Colorado.

Harrison, J.C., 1985, Earth Tides: New York, Van Nostrand Reinhold.

Helsel, D.R., and Hirsch, R.M., 1992, Statistical methods in water resources: New York, Elsevier Science Publishing, 522 p.

Laczniak, R.J., Cole, J.C., Sawyer, D.A., and Trudeau, D.A., 1996, Summary of hydrogeologic controls on ground-water flow at the Nevada Test Site: U.S. Geological Survey Water-Resources Investigations Report 96-4109. (Also available at http://pubs.usgs.gov/wri/wri964109/.)

Marine, I.W., 1975, Water level fluctuations due to earth tides in a well pumping from slightly fractured rock: Water Resources Research, v. 11, no. 1, p. 165–173.

McKee, E.H., Phelps, G.A., and Mankinen, E.A., 2001, The Silent Canyon Caldera—A three-dimensional model as part of a Pahute Mesa—Oasis Valley, Nevada, hydrogeologic model: U.S. Geological Survey Open-File Report 01-297.

Melchior, P., 1966, The Earth Tides: London, Pergamon Press.

Merritt, M.L., 2004, Estimating Hydraulic Properties of the Floridan Aquifer System by Analysis of Earth-Tide, Ocean-Tide, and Barometric Effects, Collier and Hendry Counties, Florida: U.S. Geological Survey Water-Resources Investigations Report 03-4267.

Munk, W.H., and MacDonald, G.J.F., 1960, The Rotation of the Earth: A Geophysical Discussion: London, Cambridge University Press.

Narasimhan, T.N., Kanehiro, B.Y., and Witherspoon, P.A., 1984, Interpretation of earth tide responses of three deep, confined aquifers: Journal of Geophysical Research, v. 89, no. B3, p. 1913–1924.

National Security Technologies, LLC, 2010, Completion report for the well ER-20-7, Correction Action Units 101 and 102--central and western Pahute Mesa: U.S. Department of Energy Report DOE/NV—1386.

O'Reilly, A.M., 1998, Hydrogeology and simulation of the effects of reclaimed-water application in west Orange and southeast Lake Counties, Florida: U.S. Geological Survey Water-Resources Investigations Report 97-4199.

O'Reilly, A.M., 2004, A Method for Simulating Transient Ground-Water Recharge in Deep Water-Table Settings in Central Florida by Using a Simple Water-Balance/Transfer-Function Model: U.S. Geological Survey Scientific Investigations Report 2004-5195, 49 p.

Potter, M.C., and Goldberg, Jack, 1987, Mathematical methods (2d ed.): Englewood Cliffs, N.J., Prentice-Hall, 639 p.

Prothro, L.B., and Drellack, S.L., Jr., 1997, Nature and extent of lava-flow aquifers beneath Pahute Mesa, Nevada Test Site: Bechtel Nevada, Technical Report DOE/NV-11718-156.

Rasmussen, T.C., and Crawford, L.A., 1997, Identifying and removing barometric pressure effects in confined and unconfined aquifers: Ground Water, v. 35, no. 3, p. 502–511.

Risser, D.W. and Bird, P.H., 2003, Aquifer tests and simulation of ground-water flow in Triassic sedimentary rocks near Colmar, Bucks and Montgomery Counties, Pennsylvania: U.S. Geological Survey Water-Resources Investigations Report 2003-4159.

Rorabaugh, M.I., 1964, Estimating changes in bank storage and ground-water contribution to streamflow: International Association of Scientific Hydrology, Publication 63, p. 432–441.

Sepúlveda, N., 2006, Ground-water flow model calibration program MODOPTIM and its application to a well field in Duval County, Florida: USGS Scientific Investigations Report 2005-5233. Reston, Virginia: USGS.

Stallman, R.W., 1971, Aquifer-Test Design, Observation, and Data Analysis: USGS Techniques of Water-Resources Investigations, Book 3, Chapter B1.

Theis, C.V., 1935, The relation between the lowering of the piezometric surface and the rate and duration of discharge of a well using groundwater storage: American Geophysical Union Transactions, v. 16, p. 519–524.

Toll, N.J., and Rasmussen, T.C., 2007, Removal of barometric pressure effects and earth tides from observed water levels: Ground Water, v. 45, no. 1, p. 101–105.

Walton, W. C., 2008, Upgrading aquifer test analysis: Ground Water, v. 46, p. 660–662. doi: 10.1111/j.1745-6584.2008.00442.x

Warren, R.G., Cole, G.L., and Walther, D., 2000, A structural block model for the three-dimensional geology of the Southwestern Nevada Volcanic Field: Los Alamos National Laboratory Report LA-UR-00-5866.

Weeks, E.P., 1979, Barometric fluctuations in wells tapping deep unconfined aquifers: Water Resources Research, v. 15, no. 5, p. 1167–1176.

Yeh, T.C., and Lee, C.H., 2007, Time to change the way we collect and analyze data for aquifer characterization. Technical commentary: Ground Water, v. 45, no. 2, p. 116–118.

Appendix A. SeriesSEE add-in

The SeriesSEE add-in, example data sets, and installation instructions in the zipped file, AppendixA_SeriesSEE.v.1.00.zip, can be accessed and downloaded at *http://pubs.usgs.gov/tm/tm4-F4/*. The SeriesSEE add-in, supporting modules, templates, and compiled FORTRAN codes are in the subfolder AddIN. Examples of geophysical log, data logger input, other time series, and water-level modeling data sets are in the subfolders Example_BOREHOLE, Example_LOGGER, Example_TIME, and Example_WLM, respectively. An Adobe PDF version of the help files, SeriesSEE.V1.00_Explain.pdf, is in the root directory because compressed help files that are on servers can be disabled, *http://support.microsoft.com/kb/896358*. Contents of all subdirectories are reported in README file in the root directory of the unzipped AppendixA_SeriesSEE.v.1.00.zip file.

Appendix B. Source Codes for SeriesSEE

Source code for SeriesSEE exists as FORTRAN, XML, and VBA codes in the zipped file, AppendixB_Codes-SeriesSEE.v1.00.zip, which can be accessed and downloaded at *http://pubs.usgs.gov/tm/tm4-F4/*. The FORTRAN codes NoComment and WLmodel support PEST and solve water-level models, respectively, and are in the FORTRAN subfolder. All VBA code are in the SeriesSee.V*.xlsm and SSmodule_*.xlsm files in the VBA subfolder. The XML that defines SeriesSEE commands and buttons in the Excel ribbon are in the XML subfolder. Contents of all subdirectories are reported in a README file in the root directory of the unzipped AppendixB_Codes-SeriesSEE.v1.00.zip file.

Appendix C. Verification of Analytical Solutions

Analytical solutions that were computed with the FORTRAN program WLmodel and published results of the same solutions in the zipped file, AppendixC_Verification.zip, can be accessed and downloaded at *http://pubs.usgs.gov/tm/tm4-F4/*. The analytical models for pneumatic lag, gamma, moving average, step, Theis, and tide are verified against known solutions in the subfolders AirLAG, Gamma, MovingAverage, Step, Theis, and Tide, respectively. Contents of all subdirectories are reported in a README file in the root directory of the unzipped AppendixC_Verification.zip file.

Appendix D. Hypothetical Test of Theis Transforms

The Excel program, HypoFrame, measured water levels, measured barometric changes, and reported water-level models in the zipped file, AppendixD_HypotheticalAquifer.zip, can be accessed and downloaded at *http://pubs.usgs.gov/tm/tm4-F4/*. HypoFrame is a workbook for simulating hypothetical aquifer tests and creating water levels with known pumping signals and environmental fluctuations. The premise and usage of HypoFrame are documented in the compressed help file 00_HypoFrame-HELP.chm. Measured water levels and barometric changes that serve as environmental fluctuation sources and background water levels are in the file 00_Meas+Back-for-Analysis.xlsx. Reported water-level models and tools for viewing parameter correlation are in the subfolder WLMs.

Appendix E. Pahute Mesa Example

Measured water levels, measured barometric changes, pumping signals, and reported water-level models in the zipped file, AppendixE_PahuteMesaExample.zip, can be downloaded at *http://pubs.usgs.gov/tm/tm4-F4/*. The zip file contains the pumping response in well ER-20-7 from the ER-20-8 main upper and lower aquifer tests.

www.ingramcontent.com/pod-product-compliance
Lightning Source LLC
Chambersburg PA
CBHW081409170526

45166CB00010B/3267